Page and Place:
Ongoing Compositions of Plot

Spatial Practices

An Interdisciplinary Series in Cultural History, Geography and Literature

19

Page and Place: Ongoing Compositions of Plot

Jon Anderson

Rodopi

Amsterdam - New York, NY 2014

Cover image: Iwan Bala 'In Process/Ar Waith' (2011); photographed by Toril Brancher; The National Library of Wales Collection. Also at www.iwanbala.com.

The paper on which this book is printed meets the requirements of "ISO 9706:1994, Information and documentation - Paper for documents - Requirements for permanence".

ISSN: 1871-689X
ISBN: 978-90-420-3893-6
E-book ISBN: 978-94-012-1175-8

Printed in the Netherlands

Contents

PART IV – BORDERLANDS

Author's Note

Firstly, I would like to thank all the authors who generously gave their time and insights to make this book possible. Full bibliographic references for their work are given at the end of the book. The reproduction of extracts from these authors' works has been undertaken in the spirit of fair use, but I would nevertheless like to acknowledge the following publishing houses for agreeing to reproduce extracts in this book: *Seren* (Peter Finch, Andrew Craig Williams, Richard Collins, Grahame Davies); *Parthian* (Lloyd Robson); *Vintage* (Tessa Hadley, Niall Griffiths); *Bloomsbury* (Malcolm Pryce); *Carcanet* (Gillian Clarke); *Faber* (Owen Sheers); and *Granta* (Iain Sinclair).

Secondly, I would like to acknowledge the generous funding from WISERD (The Wales Institute of Social and Economic Research) which made this monograph possible.

Thirdly, I would like to thank Iwan Bala for permission to use his 'In Process/Ar Waith' (2011; photographed by Toril Brancher; The National Library of Wales Collection), as the cover image of this book. More of Iwan's work can be found at www.iwanbala.com.

Fourthly, alongside the main text there are a number of 'boxes' that appear throughout the chapters. These boxes deliver examples to elaborate on the themes discussed in each section and augment the main argument of the book.

Finally, I would finally like to acknowledge the support of Dr Sarah Morse whose expertise in the field of Welsh literature was vital throughout the research stages of this book.

"Fiction is the lie through which we tell the truth" (Albert Camus).

"I came to value exceedingly novels ... that connected human enterprise to real and specific places, and I grew to be mildly distrustful of work that occurred in no particular place, work so cerebral and detached as to be refutable only in an argument of ideas." (Lopez, 1998:139)

"My office is in one of the streets that cower beneath the looming dominance of Cardiff Castle. In this street lived the first coal merchants and the craftsmen that filled the city with colour and light. The tribes that defended the city against invasion marched up and down outside, before they disappeared into its walls. Pryderi was born here, Torchwood was lit just a few streets away, and Shirley Bassey, that other star, is also a native of this city. Who is to say which of these is real and which a fiction?"(with apologies to Kate Atkinson, 1995:10)

"Reality is the page. Life is the word." (Mitchell, 2001:267)

Chapter 1
Crossing the breach between page and place: illuminating the relations between location and identity

Page and Place explores the geographical nature of the human condition. Geographers have long argued for the near-umbilical connection between people and place (see for example, Sack, 1997; Casey, 2001; Soja, 2010), presenting philosophical treaties and reasoned arguments for the co-constitution of humans and their geographies. This book adds to these arguments, but in a unique way. Rather than offering a scientific investigation into person-place relations, it seeks to use literature in order to gain a fresh insight into the ways in which we can understand the connections between identity and geography. Through this process, not only does this book reconceptualise the relations between literature and location – arguing that the distinctions between the fictional and the real often become entangled when we cross the breach between page and place, but also that this entanglement can offer us important insights into the nature of contemporary belonging. As we will see in this Introduction, the book calls on a range of geographical theory to contextualise the relations between people and place, before going on to explain the unique way in which this book tackles this issue. The Introduction goes on to outline the fiction used to explore these issues before summarising the key findings in the exploration of relations between people and place.

> "Literature, with its integrated triad of person, [narrative] and place, is an essential field for geography no less than for any other discipline attempting an explication of the human condition." (Pocock, 1988:87)

The fields of literature and geography have many elements in common. As Pocock tells us, both disciplines are connected by the 'integrated triad' of person, narrative and place, whilst both also seek to use these constitutive relations to offer new insights into the human condition. This book develops these premises through exploring how the imaginary worlds of stories intersect, conflict, and supplement geographical ideas to further our understanding of the people and places around us. It explores how books reshape our perception of the world and how geographical theories proffer new insights into fictional literature. In line with Pocock, the book argues that through examining the complex connections between the geography of places and the geography of pages we can gain new understandings of the human condition.

This book explores the relations between humans, stories, and places through questioning the traditional assumption of literary geography, namely, that there is a clear distinction between 'real' geographies on one hand, and 'imagined' fiction on the other. It argues that fictions are not inevitably detached and apart from real life, rather fiction becomes insinuated into the material locations of the everyday. The 'real' and the 'imagined', or 'place' and 'page', are therefore not discrete and oppositional in their relations, but rather they are thoroughly entangled. This insight is important as human beings live our lives in places. In other words, we are spatial beings. Geographical locations come to define our identity and our culture and, in turn, our lives and actions influence the meanings and significance of these key geographical sites. It is in these reciprocal relations between people and place that literature plays a key role. It is possible for literature to mediate and illuminate human life. Not only do the relations between people and place form the basis for fictional narratives, stories themselves can hold up a mirror to the real world and draw our attention to issues, problems, and ideas that escape our attention. In these ways, fiction pervades and penetrates our lives.

As humans are spatial beings, to be connected to a place is definitive of the human condition. Yet questions remain concerning the nature of these important relations. In what geographies, for example, do we sense a place of home? Through what practices do individuals develop a sense of self in place? How does this connection make us feel, and what actions does it move us to engage in? Coupled to these lines of enquiry, we can also ask questions related to the nature of the tensions that exist between our places and other places elsewhere? What threats do these other sites pose to our sense of home, and how do we react to them? What happens when our locations change, or when we no longer have a sense of place? Answers to these questions come to define the human condition and although the questions may be timeless, the way we respond to them varies in relation to different times and locations. As this introductory chapter outlines, in *Page and Place* I take my particular vantage point in time (the early decades of the twenty-first century) and place (in Wales, UK) to offer answers to these questions by engaging with contemporary geographies and fiction. By tackling these issues through crossing the breach between page and place, this book seeks to provide unique insight into the key question that defines the human condition: what is it to belong?

Places and the Page

The book starts from the assumption that fictional writing, like life, is inherently geographical in nature. As all authors are aware, fictional worlds are constituted by three key components: characters, timeline, and geographical context. Of these key constituents, geography is crucial; as Piatti et al observe, it is "impossible to even think of literature without any spatial context" (2008:4). Geography is therefore essential to fiction (see Reuschel and Hurni, 2011:291); however, the relationship between real places and literary stories is complex. In some cases, fiction intersects directly with real world cartographies. Narratives can be based in specific countries, regions, and towns, so much so that we can visit them in person and follow our characters' footsteps with our own. This direct coincidence of fictional and geographical space can be seen in examples such as Hardy's Wessex, Kerouac's California, or Auster's New York. In other fictions, real geographies are moulded, with distances reduced, streets folded and landmarks crumpled together. In this way, (brave) new worlds are invented in the author's and readers' imagination. In the same way as some authors invent 'counterfactual histories' (see Piatti and Hurni, 2009), these re-workings may be conceived of as 'counterfactual geographies'. However, as we will discuss further below, any claim to a clear and reliable reality is often difficult to maintain in the realm of literature and geography. Thus, in the words of Piatti and Hurni, stories can be rooted directly in the "physically comprehensible world", or exist in their own "rich geographical layer" above it. These complex and fascinating relations combine to produce the "geography of fiction" (Piatti & Hurni, 2011:218).

The scholarly investigation of the relations between geography and fiction has a long history; the term literary geography first appeared at the beginning of the twentieth century (with Sharp, 1904, see Piatti and Hurni (2011). However, during this history the investigation into the role that (real or imagined) places play in our stories has not always fully emphasised the complexity of the relations involved. Exploring questions such as 'where is fiction set?' or 'where are stories located?' are vital to the meaning and significance of any narrative, yet they can often underplay the significance of geography in fiction. In a similar way to a scientific framing of place (see Tilley, 1994), this perspective reduces the world around us to a passive stage on which characters make their entrances and exits. Such framing makes it easy to keep the 'physically comprehensible world' apart from the 'literary world', effectively constructing an impassable boundary that separates the 'real' from the 'imagined' (see Crawshaw and Fowler, 2008; Lahaie, 2008; Reijnders, 2010). However, this straightforward border between the real and imagined in the geography of fiction is far from impregnable. In many ways, as the quotations in the prologue imply, there are close and (con)fusing connections between the real and the fictional.

> "Geography pervades the content, practice and meaning of creative writing; it is simultaneously intrinsic to, while weaving its way into, the very interstices of the written word." (Saunders, 2010:436)

So whilst it may be tempting to keep apart the literary and the literal, it is increasingly common to "unpick the deep-rooted complexities embedded within the literary representation of space and place" (Cooper & Gregory, 2011:91). As Saunders tells us, geography pervades creative writing, whilst creative writing also penetrates geographies. As a consequence, it makes little sense to erect a border between the worlds of 'fiction' and 'reality', rather it is better acknowledge the existence of a 'permeable membrane' (after Piatti, et al., 2008:5) between these worlds that can be crossed at will. If we accept this position, our starting question in exploring the geographies of fiction should not be 'where is fiction located', but rather, 'where does fiction *take place*'?

How literature takes geographies: Entangling the relations between 'page' and 'place'

The notion of 'taking place' confers two significant meanings for the relations between literature and geography. The first meaning aligns with the traditional approach to literary geography – a story, or perhaps more particularly an event within a narrative, straightforwardly occurred; it happened or took place. In this sense, the location had little effect or influence upon the event, and the event itself did little to change the geography in which it occurred. The second meaning, however, goes further than this invocation of the 'simple location of things' (after Casey, 2005). It suggests an agency and consequence to this event: the event *took place;* it aggressively, passively, intentionally, or otherwise, took a place for a given length of time. As geographical locations come by their meanings as a result of the complex intersections of events and practices (see Anderson, 2010), this *taking place* has an effect on the location in which it occurred: it alters that place's use, identity, and meaning. From this perspective, literature takes and makes places alongside other actors within the cultural world; it becomes one contributor to the meaning and identity of the places around us. Geography is thus not only essential to fiction in terms of grounding, locating and staging a narrative, but fiction in turn actively creates geographies.

The role that literature plays in actively illuminating and mediating the relations between people and place means that in order to investigate the geography of fiction we must do more than codify the simple location of stories. Instead we must be sensitive to the ways in which the permeable membrane between literature and geography is continually crossed by both the writer and the reader (as we will discuss further below). In short, we must diagnose and explore the reciprocal relations between page and place. In this reading, fiction is actively cartographical, fashioning and refashioning the world around us. It simplifies, highlights, and draws our attention to the strange or the familiar. Fiction therefore can be considered a form of *geographical cartography*. As Tally (2011) argues,

> "...Narrative itself is a form of mapping, organizing the data of life into recognizable patterns with it understood that the result is a fiction, a mere representation of space and place, whose function is to help the viewer or mapmaker, like the reader or writer, make sense of the world".

Through its composition, fiction alters our perception and understanding of the world, producing new versions of reality. As Price tells us, fictional narrative,

> "allows the flow of time-space to be arrested and shaped in ways that make deliberate sense of events' interlinkages to one another and to a

> broader trajectory of meaning, and that allows us to emplace ourselves in that flow of events in meaningful ways." (Price 2013:122)

If we activate the previously passive relations between literature and geography in this way, we become sensitive to the agency and influence that fiction has on place, and vice versa. We can no longer maintain a rigid separation between the literary and the literal, but rather need to examine how the permeable membrane between them is crossed and contaminated. *Page and Place* uses the spatial metaphor of 'entanglements' (following Sharp et al., 2000, and Foucault 1979) to conceptualise these crossings, suggesting how the literary plot of the story and the literal plot of a location are always implicated in, and constitutive of, one another. To paraphrase Sharp et al., the:

> "use of the term 'entanglements', suggesting an image of knotted threads, is intended to underline the deep 'spatiality' of this spinning together of [in this case, page and place]. We talk about 'entanglements' to indicate that [page and place are] *always* played out in [and] across [each other]." (2000:1, my emphasis)

Framing the literary and the literal as entangled phenomena suggests, therefore, that page and place cannot exist independently of each other. This book argues that the literary relies on the literal for a grounded materiality, whilst the literal relies on the imagined for an alternative reality, for escape and for hope. The literary and the literal are thus hybrid phenomena, existing interdependently. As Sharp states, "the one always contain[s] the seeds of the other, the one always bear[s] at least a trace of the other that contaminates or subverts it" (2000:20). The literary is thus always present in the constitution of the literal, and the literal is always present in the constitution of the literary. There is no complete separation between page and place.

From this entangled perspective, it makes more sense to suggest that we are not dealing here with page *and* place, but page *in* place (and place *in* page). Akin to a Foucauldian understanding of power (Foucault, 1979), page and place circulate within and between one another. Borrowing language from geographical theory, there is a 'constitutive coingredience' of page and place (after Casey, 2001, and see below). The page and the place are always connected; storylines can take you to an actual territory, whilst a real place is constantly shadowed by our literary emotions and imagination. Such a claim is explicit when a story or novel is based in a real time and place (as suggested by Hardy's Wessex, above), and may be more implicit in fantasy or science fiction. Yet despite this, the allegorical intentions of the author and the imaginations of the reader often compress distance, time, and social context and, in the turning of the page, enfold the literary into the literal, imagining both presences in each world (be it ourselves in an-other's world, or an alien world co-inhabiting ours). Thus through wormholes and portals, wardrobes and changing rooms, the literary can facilitate escape whilst at once bringing us home to familiar territory.

Capturing page and place: the notion of Plot

If we take the hybrid nature of page and place seriously, this book suggests we can capture this idea of entanglement through expanding upon the notion of 'plot'. Plot can be understood in both literary and literal senses. Plot is at once the crucial narrative or story line of creative writing (as introduced by Pocock, above), whilst also being a locatable, geographical territory – for example *a plot of land, this blessed plot, this sceptred isle* (after Shakespeare, Richard II, Act II, Sc1). The entangled nature of these different, yet connected, aspects of plot is indicated by a third definition – the ability to intrigue, subvert, scheme, or imagine. Just as we think we have pinned the plot as 'simply' a story telling device, or a geographical location, it loosens its independent moorings and re-tangles its-'self' with its 'other' – our plots are therefore never singular in meaning, but always plural.

From this view, there is no distinguishable 'real' from 'imagined', no independent page from stand-alone place, or discrete plot (in the sense of fictional narrative) from discrete plot (in the sense of location). The imagined versus reality dichotomy is therefore thoroughly flawed. As Aitken & Dixon suggest, we can no longer talk of the creative process, "representing, or mimicking [an independent] reality, because we can no longer assume that there is a single, coherent reality waiting out there" (2006:327). Literary genres may create their own worlds and their own 'reality-effects', but these are entangled within our inhabited, literal places. As Reijnders suggests, it makes little sense to talk of two separate worlds (of the imagination on one hand, and the real on the other), but rather of a, "complex of multiple imaginations and realities" (2010:41).

It is here, therefore, that the entangled nature of plots becomes vital. This book argues that in this complex of multiple imaginations and realities, places are epistemologically involved in the fictions we read and write, and fictions are active in constructing the geographies around us. Place on the page is thus not a sterile stage for human action, rather place contributes to the action; as Piatti & Hurni put it: "Geography never becomes a simple backdrop within the fiction; the different settings in the quarters have almost protagonistic qualities" (2009:340). Place and space thus emerge as more than simply a "background complication" (Soja, 2010:19) for literary geography, instead locations become framed as an actor themselves, both contributing to the story, and playing its part on the page. The geographical plot is thus a key driver of the fictional plot.

Framing the relations between page and place in this way is informed by, and indeed further revolves, the 'spatial turn' that has occurred in the social sciences and humanities over recent years. This spatial turn involves recourse to geographical theory in order to further understand the role of place in our lives. This turn has set in motion many avenues for inquiry across disciplines, however two key shifts in academic understanding can be seen to affect us directly here: the 'mobilisation' of place as a concept, and the 'death of the author' narrative. The next sections outline, firstly, a brief introduction to the spatial turn, and secondly the consequences of these two shifts for how we think about the relations between people, fiction, and place.

Spatial turns and writing places: geographical theory and the page

> "The 1980s and 1990s can now be recognized as marking a distinct and influential 'spatial turn' in the humanities and social sciences, as cultural geography embraced social theory to explore 'the created space of social organization and production', the 'concrete' and socially 'embodied' forms of spatiality, and the navigation of both 'physical and mental space'." (Walford Davies, 2012:3)

The spatial turn across the social sciences and humanities summarises the ways in which a wide variety of scholars have begun to acknowledge that the geographical dimension of our lives "has never been of greater practical and political significance" (Soja, 1996:1). By drawing on an eclectic range of theoretical positions, scholars have begun to frame the world as both social *and* spatial in nature, beginning to explore how places play a key role in making us who we are (see Anderson, 2010). The spatial turn has facilitated insights into how human identity and culture is indelibly connected to the places and spaces around us. It has shown how we are defined to differing degrees by these places and how in turn we define these geographies. In sum, as Soja tells us, the spatial turn has emphasised how:

> "Human life is.... spatial, temporal, and social, simultaneously and interactively real and imagined. As intrinsically spatial beings ...we are at all times engaged and enmeshed in shaping our socialised spatialities and, simultaneously, being shaped by them." (Soja, 2010: 18)

The spatial turn has led to a number of "profound and sustained transformations" (Warf & Arias, 2008:2) in the way places have been understood by scholars. In line with the increasingly interdisciplinary focus on place, sites are now rarely considered solely in terms of their narrowly-defined geographical dimensions, but rather as inherently connected to time, space and society. This growing acknowledgement has changed the study of place from one that was seen to be "highly abstract and remote from experience" (Tuan, 1974: 151) and towards an approach that considers place as "thoroughly meshed" (Casey, 2001: 684) into the human condition. As Entrikin (2001:694) puts it:

> "for most of the past century, [scholars of space] have approached their subject in a manner that could be characterised in terms similar to those used by Italo Calvino [1986] to describe the fictional citizens of Baucis, who, as residents of a city built on stilts, lived in the sky and gazed at the earth through telescopes, never tiring 'of examining it, leaf by leaf, stone by stone, ant by ant, contemplating with fascination their own absence'."

Scholars within the spatial turn have challenged this distancing of people from place and rejected conventions that reduce space to an inert background and time to a fecundity (see Crang and Travlou, 2001:163). These insights stem in part from existential accounts of 'Being-in-the-World' (from Heidegger, 1988) and the exploration of perception through phenomenology (from Merleau-Ponty, 1962). They have also been extended through geographical adventures "in the dialectic between people and places" (after Davidson, 2003) recorded by Tuan (1974), Relph (1976), Seamon and Mugerauer

(1985), Sack (1997), Hillier (2001) and Preston (2003). In general, these works posit that the human condition is a profoundly geographical one, with identity both influencing and being influenced by its inhabited material places. As a consequence, these traditions recognize geographical context and the passage of time in place as integral to human existence. From these perspectives places are not passive backdrops on which actions occur, rather they are the medium that impinges on, structures and facilitates these processes. As Tilley states, places are "involved in the action and cannot be divorced from it" (1994:10). Places then, are not only a medium but also an outcome of action, producing and being produced through human practice. It follows then that through inhabitation and dwelling, a person–place relationship is inevitably developed. As Casey (2001) suggests – there is no self without place, and no place without self.

As the medium and outcome of personalized events and activities, human memory and identity become bound to places by different degrees. As a result, as Casey points out, "places [can] possess us – in perception, as in memory . . . insinuating themselves into our lives" (Casey 2000:199). I argue here that the same process of co-constitution – or entanglement – occurs between the fictional page, human lives and geographical places; in other words, through the active reading of literary and literal plots. As with human events, stories become sedimented into the landscape, with fiction and geographies helping to construct and reproduce each other. Places, like people, have biographies in as much as they are formed, used and transformed in relation to practice. Places help us to recall stories associated with them, whilst stories acquire part of their mythic value and historical relevance when they are rooted in the concrete details of locations in the landscape. Places thus not only impinge upon, structure and facilitate the literary, but the literary is also involved in our constructions and lived experiences of literal places.

The spatial turn has thus resulted in key transformations in understanding the relations between people and place. However, as part of this broad turn towards recognising the importance of the spatial in our lives, two key debates serve to further illuminate the interconnections between fiction and geography, namely the 'mobilisation' of place as a concept, and the 'death of the author' narrative.

From place as a 'constant' to place as an 'ongoing composition'

Traditionally, geographical sites have been framed as fixed and static in nature (for a full review see Anderson, 2010; and Massey, 2005). Places are, of course, static in terms of their grid reference and location. However, this approach to place also considers sites to be static in the sense of being changeless, preserved, and constant. Places may grow or contract, but remain more (or less) of the same. From this view, if places are 'carved out of space' (Sack, 2004:244), then it makes modern sense that they come to be defined when such processes occur over a long period, stabilising and making durable place identities which become widely recognised and acknowledged. As Tuan argues, from this perspective place becomes an:

> "essentially... static concept. If we see the world as process, constantly changing, we would not be able to develop any sense of place" (1977:179),

and as Relph offers, it is every geographical site's,

> "persistent sameness and unity which allows [it] to be differentiated from others." (Relph 1976:45)

From this perspective, therefore, the process of place composition involves the "carving out of '*permanences*'" (Harvey, 1996:294). In the words of Dovey (2010:3), places come to be, "identified with what does not change; their 'sense of place', 'character' or 'identity' is seen as relatively stable" (Dovey, 2010:3). Such an approach to place thus formulates a world of closed, bounded and stabilised sites. Place identities and the cultures that create them are deemed to be coherent and preserved, with any narrative of change being framed around the threat of 'foreign' ideas, peoples and commodities, "rupturing and contaminating" local places and values (Coleman & Crang, 2002:1).

As Price has told us (above), fiction can play a key role in perpetuating a static sense of place. In her words, "narrative is a primary approach to the carving out of the 'permanences' invoked by Harvey to define the act of place-making" (2013:122). Fiction can achieve this through arresting the passage of time and fossilising a place in a particular era or period. Fiction can also idealise a rooted and static formulation of place through the ideas proposed in a narrative. Erhlich, for example, suggests such a prescriptive ideology, commenting that our lives should not be:

> "a series of dramatic events for which he or she is applauded or exiled but a slow accumulation of days, seasons, years, fleshed out by the generational weight of one's family and *anchored by a land-bound sense of place.*" (1985:5, my emphasis)

In the same way, as Alexander as noted, the poet Seamus Heaney often recommends an "essentially sedentary understanding of place and belonging" in order to frame the geography of fiction (2010:11). In Alexander's words:

> "Heaney's affirmation of the importance of a direct, perhaps primal, connection with 'the land itself' ... compounds being and dwelling in a conception of place as stable, fixed, and intimately familiar...." (2010:37)

As Hewitt confirms, this perspective argues that a writer,

> "must be a rooted man [sic], must carry the native tang of his idiom like the native dust on his sleeve; otherwise he is an airy internationalist, thistledown, a twig in a stream. [The writer] must have a native place, pinpointed on a map, even if it is only to run away from." (cited in Alexander, 2010:37)

The pull of a rooted and static sense of place is thus widespread in the geography of literature. However, with the onset of relational approaches within the social sciences in general (see, for example, Jones, 2009) it is also possible to frame place in new, challenging, ways. These relational framings argue that place need not be approached as sedentary and rooted, but rather should be understood as unstable, mobile, and *routed* in nature.

> "The concept of place has, over the past decade, been invigorated theoretically by geographers emphasizing the unboundedness, historical dynamism, and multiple identities inherent in places. This work is often characterized as a new way of conceiving place, enabled in part by the rise of postmodern cultural and social theory and the related demise of modernism in academia." (Oakes, 1997:509)

Influenced by the spatial turn and the theoretical impetus from other disciplines, the notion of place has changed from one that is sedentary and stable to many that are provisional and emergent. This approach rejects a static ontology of 'being-in-the-world' (associated with Heidegger, 1956), and embraces a more emerging ontology of 'becoming-in-the-world' (associated with Deleuze, 1985, 1993, see Dovey, 2010, and Anderson, 2012). In respect to places, therefore, such an approach insists that geographical sites are not "stabilised, normalised, sedimented or structured [rather] they are always in a process of dynamic unfolding and becoming" (Rose, 2002: 385). In the words of Doel, "place is nothing if it is not *in process*" (1999:7). From this perspective, place is not a given, but something immanent, forever forming and in progress; as Dovey states, "all places are in a state of continuous change" (2010:3). Therefore this approach emphasises the changing rather than the constant nature of place. Place is not constant or static in any real sense, but dynamic and active. This conceptualisation also alerts us to how place need not be defined simply by *boundaries*. Instead, place can be defined by comings together or connections from elsewhere and elsewhen (see Massey, 1993). From this perspective, therefore, place is far from a static, stable or fixed entity. Place is no longer reliable, consistent or necessarily coherent, it is wholly provisional and unstable. Place, at any moment, emerges in time and space from the web of flows and connections meeting at a particular node. Place becomes the 'point' in this motion, a temporary pause. However, this pause is not solely geographical but also temporal, a place is *now*, it is permanently in the present, only temporarily 'fixed', and now something else.

For Ingold (2005) these coming togethers are not static or 'finished objects', rather places are 'dots' (i.e. pauses) in '*trails of action*'. For Ingold, when configuring place it is crucial to remember the movement that brought various components together at a geographical site, where they came from, and where they are going next. In this view, the *meeting* (see Massey, above) cannot be separated from the *movement*, the 'dot' cannot be separated from the 'continuous line' (Ingold, 2005). In Ingold's words, even though:

> "we are now more inclined to think of [place] as a complex of interconnected *points* [it is more useful to think of place as a complex of] interwoven *lines*. ...The lines of the meshwork... are the trails along which life is lived. And it is in the entanglement of lines, not in the connecting of points, that the mesh is constituted." (2005:47, my emphasis)

I will return to the notion that are ideas of place are 'trails' or 'lines' below, but as we can see here, places have been thoroughly reconsidered from entities that are static and rooted, to entangled processes on "on-going trajectories" (Massey, 2006a:92). Here place has become a moment and a location, simultaneously shot through with lines of movement that constantly (re)combine to change its form and substance. In geographies of fiction, this approach can be initially identified in writers such as Charlotte Williams who talk of places (and their cultures) being in a state of constant becoming. Rather than being fixed and monumental (like unread books), Williams suggests places are better conceived as processes that are performed in the here and now (in the same way as stories are when they are read and shared); as she states:

> "How could anyone think a [place] stands still? Or that we hand it over intact like a book? It's more like the retelling of a story. It changes every time we put our own spin on it and pass it on. We add our own little bits, forget others and get some of the story completely mixed up." (2002:182)

Thinking about place in this dynamic way is an important shift. It has the effect of not only countering the inertia of the static and conserved, but also of transferring agency to the contemporary inhabitants of locations, to those people who take and make their histories and geographies in the present (see also Hones, 2008). Acknowledging the view that "every place is in constant change" (Solnit, 2010:2) means that place as a concept (and a reality) is no longer reduced to the discrete and essential, but rather has become relational and multi-vocal in nature. Following this approach, scholars argue that there can no longer be one true version of any place, insisting that all places "continually resist a[ny] meta-narrative" (after Pile & Thrift, 2000:303). This resistance to meta-narrative is not only due to the rise in viewing places as ongoing compositions, but also to the acknowledgement that places are 'actor-centred' processes too (after Jones, 2009). This latter claim suggests that every (real and imagined) inhabitant of a place will have a different perspective upon it, and as a consequence, a place's identity changes from the standpoint of each individual. Understanding this latter claim is particularly important when we consider the second key consequence of the spatial turn that is relevant to the geographies of fiction: the death of the author.

Engaging the reader in the composition of plot

It is clear that the author of any story is central to composing the geography of fiction. It is commonly accepted that authors write from somewhere, from what they know, and where they gained that knowledge. Authors may be inspired by a place or want to recapture the emotions a place contributed to. Through their writing they attempt to convey this knowledge and emotion, and, to some extent at least, their own preoccupations, desires and anxieties about the human condition (after Alvarez, 2009). In this way (as we have seen above), authors make their own version of a place and through their writing they create their own cartographies.[1] Like all map-makers, writers thus present their own version of the geographies around us by fashioning a positioned and selectively rendered composition of a complex reality.

However, we should not be solely interested in the geography of authorial intent, but also the geography of reception (after Sharp, in Johnson, 2004:92). Just as the author is an active contributor to the plot, so too is the reader. Following Barthes and his seminal argument 'The Death of the Author' (1977), the reader is no longer a passive consumer of authorial intent; rather they/we have the agency to produce their/our own reading of any story. Indeed as Ljungberg argues, "engaging the reader in the production of meaning...is the very essence of literature" (2003:174, also cited in Piatti & Hurni, 2009:340)). Thus the active reader brings their own emplaced knowledges to the page, and as a consequence, there is no longer a sole original author composing the plot, but a number of authors, including the reader (see also Knights & Thurgar-Dawson, 2011). As Tally (2011:no page) tells us, "... As writers map their worlds, so [too do] readers or critics ... in order to orient ourselves and make sense of

---

1 Any fiction, therefore, functions in a similar way to a map. As Harvey (1969, 1972) tells us (see also Hubbard 2002:3), maps function as partial portrayals of complex landscapes. They reveal relationships and patterns, and construct languages to store and communicate understandings of the world. Literature functions in a similar way, as Tally (2011:no page) elaborates:

> "...storytelling is an essential form of mapping, of orienting oneself and one's readers in space. The storyteller, like the mapmaker, determines the boundaries of the space to be represented, selects the elements to be included, establishes the scope and scale, and so on. In producing the narrative, the writer also produces a map of the space, connecting the reader to a totality formed by the narrative itself. Narrative is thus a spatially symbolic act in establishing a literary cartography for the reader."

things in a changing world". For some readers the original author's intent *may* be read directly, but it may also be missed, read aberrantly or trangressively (see Hall, 1997). For every reader, therefore, the plot becomes their own. In this way the literary plot is "eternally [re]written here and now" (Barthes, 1977:4), and, with each re-reading, the consumer of the text become an active contributor to its diegetic meaning.

From this perspective, books and the places they relate to are far from being 'stable ontic entities' (after Kitchin & Dodge, 2007:341, cited in Hornsey, 2012:675), but are better framed as 'fundamentally ontogenetic' in nature. To paraphrase Kitchin and Dodge, these entangled plots are, "always in a state of perpetual becoming and constantly remade via a dispersed set of encounters between the printed artefact [and], the particular user" (ibid.). Johnson acknowledges the impact of this entanglement in authorship with respect to her study of Dublin's fiction. She suggests that a diversity of geographies are produced as each reader creates their own entanglement of plot; as she states:

> "there is not a single map of the city but a variety of journeys—material, existential and metaphorical ...As Lyons tellingly notes, 'There was not one Dublin, but several different Dublins and how it looked depended on the eye of the beholder'." (2004:98/100)

Taking the insights from the spatial turn, the mobilisation of place as a concept, and the death of the author narrative together, this book argues that fiction does not exist solely in the world of the imagined or the world of the real, but is entangled in both. The entanglement of page and plot is not a static or absolute stable 'thing' (although some entanglements will be more durable than others); rather it is better understood as an ongoing process. Plots are *composed* – they have a position (in time and place), and are performed by numerous actors (the author of the book, its reader, and others that contribute to the materialities and meanings of place).[2] As a consequence, there is not

---

[2] Plots are also *de*composed – they are dynamic, they move (on), erode or diverge. They may stay in our memory as artefacts or reminiscences, conjoined with particular (re)incarnations of a place; but they also develop and grow, becoming something new as our experiences and life in plots move on.

simply one plot, one narrative, or one geographical place; rather, to paraphrase Pinder:

> "it is the very condition of the [plot] to be plural, with a multiplicity of stories, an inexhaustibility of narratives, peopled with strangers and difference." (2001:3)

Any page and any location is therefore best described as many worlds in one place (after Solnit, 2010:VII). As Griffiths puts it (see Chapter 6), it is as if stories and places "suppl[y] the materials, and what you build with them is entirely up to you" (2008:11-12). Our own plots are therefore 'what we make them', they are a jumbled combination of ingredients "waiting to be assembled' (Griffiths, 2008:12).

The geographies of fiction: from objective cartography to the composition of 'plotlines'

We have seen, therefore, that if we take seriously the spatial turn, the mobilisation of places, and the death of the author narrative, it raises significant questions for any attempt to fix and map the geography of fiction. As Solnit tells us, the geography of fiction cannot be,

> "delineated either by fences on the ground or by boundaries in the imagination. ...Something is always coming from elsewhere. [This geography] exists... but it exists in relation, whether symbiotic with or sanctuary from the larger world." (Solnit, 2010:VII)

As a consequence of this multiplicity of mobility and composition, traditional cartography cannot objectively and essentially capture the geography of fiction (see Reuschel & Hurni, 2011); as Solnit emphasises:

> "Every place is if not infinite then practically inexhaustible, and no quantity of maps will allow the distance to be completely traversed." (2010:2)

In order to use the geography of fiction to explore the relations between people and place, this book introduces the notion of the 'plotline' to acknowledge the infinite variety involved in every reading of page and place. Drawing on Ingold (above), it does not seek to map or locate a *complete* geography of fiction, as this goal is too multiple, dynamic, and forever now composing to ever fully capture. However, it does seek to create one route or 'trail' along which the geography of literature has been 'lived' and experienced (after Ingold, 2005:47). This route is called the 'plotline' through the entanglement of page and place.

Plotlines are particular routes chosen through the entanglement of fiction and geography. These plotlines are both located and literal, referring directly to a position on the globe but also inevitably entangled with fictional plot and narrative. Knotted together with these components are the lives and imaginations of the reader, who bring their own outlook, culture and ideas to the plotline, making it actor-centred and personal whilst retaining the potential to offer a collective 'way' through the geography of fiction. In this sense plotlines can be compared to Aboriginal 'songlines' (as described by Chatwin, 1987). Although songlines are inherited through family and ancestry, plotlines are democratic and personal in nature – any reader can 'call forth' a new world through their reading of a fiction. However, both plot- and song-lines create new worlds through the act of creatively reading or narrating worlds into being. Plotlines therefore exist through practice; once stories lay untold or unread, their worlds decompose and die. Plotlines thus emphasise how reading can be a creative act, bringing into being new routes of excitement, intrigue, and experience to the geographies of fiction. These plotlines therefore leave new 'traces' over this cartography (see Anderson, 2010), adding new trails of words and ideas that can themselves lead to new experiences.

If imagined and practiced in this way, individual plotlines function in a similar way to both literature and geographical theory. They function as 'ways' of communication (after Chatwin, 1987) between different people, ages, and locations. Although not definitive or absolute, they remain a partial and selective 'map' and 'history', functioning as both a 'library' and a 'museum' for a neighbourhood, city, or region. In this way, plotlines can be read, followed, and mapped for a whole country, and that whole country can itself be known via a maze of plotlines. However, as plotlines are not definitive or objective, their cartographical alignment with real places may remain tentative and provisional. For some, their connection to 'real' places may be difficult to grasp scientifically, but in terms of personal affinity, poetic resonance, or lateral memory, they offer new entangled co-ordinates in the geography of fiction. In this way, plotlines are akin to memory myths (Harrison, 1986) or 'mythogeographies' (Smith,

2011): plotlines may, "select, distort and compress [their] subject matter" and so may "not stand examination in the light of strict [geographical or] biographical truth" (Harrison, 1986:119), however they,

> "do bear a relationship to this truth, albeit an oblique one, and they possess a truth of their own, a metaphorical truth [which] digests and makes sense of [our lives and] provides a way of giving significant shape to the sequence of events that make up the development of self." (Ibid.)

Although plotlines are ongoing and actor-centred, they nevertheless offer us the possibility of new insight into the relations between people, place, and fiction. As one record of the journey within page and place, a plotline can be cartographically recorded, not perhaps in the sense of traditional mapping, but as Smith states, in a way,

> "that the meanings ...it sets in motion can be taken from the work and applied to the street and the city as an analysis as well as an aesthetic provocation." (2011:266)

Following Smith, plotlines therefore become "exemplary and potentially realisable" maps through the world (see also de Certeau, 1984, and Bryden, 2012:213), and *Page and Place* offers a series of evocative plotlines to demonstrate how the literary is rooted in the real, and how the literary further composes our geographies. To this end, *Page and Place* focuses its plotlines in one geographical area. This book Maps eleven plotlines of the geographies of fiction in Wales. The next section explains the relevance of this geographical focus, before going on to explain how the plotlines were generated in practice.

Ongoing compositions of plot: plotlines of Wales

> "To say that the Celts of Britain have a close affinity with the land …is to underplay a near-umbilical connection. The huge skies of the Atlantic shore, the rose and gold of sunsets dying in the west, the bleakness … and the majesty of land… pierce the hearts of those who are born there. Even if they leave, the west never leaves them." (Moffat, cited in Griffiths, 2008:29)

As Moffat romantically outlines, Wales as a nation and as a people has a strong affinity to the land. This affinity resonates through much of its fiction. Many of the nation's writers display sensitivity to the relations between stories, humans, and place, and due to this sensitivity alone, Welsh writing in English provides a rich resource through which to explore the relations between people and place in fiction. However, focusing on Wales for *Page and Place* is justified not simply due to this historical affinity between people and place, but also due to contemporary pressures which locate Wales in a unique position in the early decades of the twenty-first century. Waves of colonisation, initially by the English state, and now by forces of economic and cultural globalisation, seek to change and perhaps loosen historical ties between the Welsh landscape and its people. Whilst these contemporary pressures test historical roots, the coming of devolution and the creation of a Welsh Government offers external confirmation of statehood for Wales within the United Kingdom. Allied to this, the Welsh language now enjoys equal status alongside English as the official language of public bodies, and growing (but geographically uneven) use in everyday conversation. Due to these conflicting forces, in many senses contemporary Wales can be seen to be a 'Janus' state in its contemporary form (after Garton Ash, 2005). Like the Roman god of bridges and thresholds, Wales can be thought of as having two 'faces'. One of Wales' Janus faces looks inward and to the past, whilst the other looks outward and to the future. This 'schizophrenic' nature of Welsh identity (see Finch, Chapter 2) neatly summarises the competing tensions between people and place in the twenty-first century.

Janus Wales

As we have seen, one Janus face of Wales looks to the past. This face acknowledges and celebrates the historical affinity to landscape that originated in Welsh agricultural and mining economies, and subsequently became definitive of the oral traditions of the chapel, club, and family conversation (as Box 1.1 explains in more detail).

Box 1.1. Facing the past

Many contemporary Welsh writers were brought up within communities which celebrated place as a vital component of identity. As National Poet Gillian Clarke describes (see below and also Chapter 9), there are family traditions in Wales which recite and site national mythologies in familiar landscapes and times:

> "My father set the stories of the *Mabinogi* in locations wherever we happened to be - in Fforest, in Cyncoed, in Kingsway, in Roath Park. ... these were the first stories located in a geography that was mine." (2008:8)

This intertwining of local places into national legends is common in Wales, as the fictional heroine Sarah recalls in Owen Sheers' novel *Resistance* (see also Chapter 11):

> "'[I'll tell you a] story about that hill,' he said, pointing his brush at Y Twmpa. 'About all these hills.' He swept the brush wider as if the real hills were his canvas now. He turned to look at her again. 'A story about your ancestors.'
>
> And that was when the poet began to tell Sarah his stories, recasting the land and hills she'd known all her life as the backdrop for his Celtic myths, for tales of saints and soldiers, of kings and bards. His stories worked upon the valleys around them like his paintings. He spoke of places she knew or that she'd heard of before, St Peter's Well, The Abbey, The Cat's Back, St David's Cell, but the lens of his stories made them all new again. Some of the stories she'd even heard before, but never like this, never growing from the very hills of her birthplace. Stories of King Arthur, Lancelot and Percival. About Welsh princes, Irish princesses and English armies." (2007:173)

Contemporary versions of this personalisation and localisation of stories also occur in televised drama (be it 'Torchwood' (see Chapter 2), 'Dr Who', or 'Sherlock' in Cardiff, for example, or 'The Passion' in Port Talbot (see Chapter 11)). Alongside narrative fictions these dramas provide further evidence of the continued importance of storytelling and the role it plays in sedimenting meaning into the landscape and creating ties between Welsh communities and their local places. These entanglements are thus imagined, written and told in the stories formed by and through Welsh culture, this entangled experience is "mediated, modified and deepened by literature" (Clarke, 2008:77). In *A Phenomenology of Landscape,* Christopher Tilley notes this point, he states:

> "[W]hen a story becomes sedimented into the landscape, the story and the place dialectically help to construct and reproduce each other. Places help to recall stories that are associated with them, and places only exist as (named) locales by virtue of their emplotment in a narrative." (1994:33)

Writer and poet Owen Sheers agrees, suggesting that stories offer:

> "a double-sensation of recollection and illumination.., an experience ... which expands my existing knowledge of a place to change the way I will see and experience that place, or a type of place, for ever." (2009:III)

In practice, this connection between stories and places means that local landscapes act as a memory of culture, a material cue to prompt the recollection of past events that (continue to) define the people and their place. In Clarke's words,

> "the features of the landscape are a mode of communion with a something other than themselves, a something to which we ourselves still feel we might belong." (2009:132)

Rooting such stories in familiar landscapes offers individuals the right to recognise their local place as valuable, as well as confirming the vitality of story-telling as a means to imagine and conceive the relations between people and place, and the past and future. As Clarke concludes:

> "...My father's version of the tale[s] gave me possession of what was rightly mine, and should belong to all children. It offered me a place in the myth, and gave myth and naming a place in my imagination." (2008:11)

The 'inward looking' Janus face of Wales is protective of the role that place plays in forming identity. It also remembers a succession of invasions that sought to silence local cultures and impose alien traditions in the landscape. In terms of the landscape, decisions made by UK governments in the twentieth century have meant dramatic incursions onto the physical geography of Wales. Conservative government decisions to tackle Trade Unions and switch to offshore and overseas power sources had huge effects on coal mining communities (see Davies, 1994), whilst strategic decisions to increase the water supply to English cities meant the destruction of North Walian towns through permanent flooding.[3] Griffiths (2008) outlines how the growing presence of wind turbines, imposed in his view by a 'foreign' power, is a contemporary threat to the connection between the Welsh and the land:

> "[Turbines are another example of the] exploitative attitudes of a colonial power towards its territories, and whilst the turbines don't pose the same immediate physical danger as say, the slag heap did above Aberfan in the 1960s, they're still of a piece with policies which are, at best, diffident and, at worse, criminally irresponsible; blight the hills above Chequers or Sandringham with these spindly, ugly, dominant things? Monoculture-ise the hills around Balmoral with one tree specie only? Allow a teetering mound of sodden slag to grow above Eton or Harrow?
>
> Jim Perrin ... contributes regularly to the *Guardian*'s 'Country Diary' column, and in March of 2007 he wrote about camping on Pumlumon Fawr and waking to see 'hundreds of wind turbines' on the ridges. 'The MSPs,' he writes, 'should come here to see how completely these presences destroy the sense of wildness that is crucial to our wellbeing. You can't argue with its proponent's basic intentions, but wind technology is inappropriate for Britain's last wild places – a plain moral fact about a crucial spiritual resource that governments and 'green' zealots must heed before the land is entirely lost'." (Griffiths, 2008:151)

The strength of Welsh feeling for the land and against colonial acts of incursion has led to protests which Griffiths has fictionalised to gothic degrees (see Chapter 6) and Gillian Clarke has captured in her poetry (see Chapter 9). These acts of protest and storytelling combine to further bind people to place through adding pangs of nostalgia, lost wildness, and a wish for autonomy into narratives and culture. So

---

3 As Griffiths outlines:

> "Liverpool could've easily been irrigated by other, already existing reservoirs. It was a despicable, criminal act of colonialist aggression, and [it remains] painfully poignant. Lost homes. Longing. Anger and aching." (Griffiths, 2008:29)

much so, that as Richard Collins (Chapter 7) puts it, for many Welsh writers,

> "there are times in your life when the past weighs heavily and the landscape is laden with memories, [at these times] you're living as much in the past as you are in the present." (Collins, Personal Communication (PC))

In terms of language, the historical silencing of Welsh words is particularly significant. This is due to the unique qualities of *Cymraeg* to articulate the co-ingredient relations between people and place. As Pearson and Shanks suggest, the Welsh language "is folded in the land" (2001:138), and of course, vice versa. For example, the Welsh words *tirwedd* and *tirlun* reflect the sensitivity of Cymraeg to different lived experiences of landscape which have no direct equivalent in English. *Tirwedd* describes the land's shape, form, and physicality, whilst *tirlun* reflects the image of landscape and our perception of landscape in social and cultural terms (in terms of aesthetic appreciation and notions of beauty, for example). Sensitivity to the importance of the human and cultural involvement in place is also present in the Welsh concept of *cynefin*. As Pearson & Shanks tell us,

> "cynefin is more than landscape and scenery. It is a piece of earth where a community has lived – a community with whom we identify. In this bond, language has its essential place, and here again the local factor is to the fore. The language of each bro [or area] has a distinctive hue ... a storehouse of the transmitted legacies of experiences and imaginative constructions of those particular parts." (2001:139)

As the Welsh language reveals this duality of the landscape concept, it may follow that the awareness of landscape as cultural construct is more apparent in Welsh culture (both in the Welsh and English language articulation of them). As Clarke states, "nothing is until it has a word" (2008:9). The existence of such sensitive vocabulary firstly legitimises, then in turn affirms and encourages, the importance of the relations between people and place. If these words disappear, it follows that the relations cannot be articulated or linguistically communicated, and communities cannot be connected and mobilised around them (see Box 1.2). Although in the twenty-first century the Welsh language has been reinstated as a national language (see Chapter 10) many individuals remain sensitive to the threat posed by the monoglot of English words and ideas.

Box 1.2 Hiraeth and home

As we have seen, in both Welsh language and Welsh culture there is a sensitivity to the connections between people and place. This is not only articulated in its language, but this language further entangles the co-ingredient constituents of geography, culture, and story. Notions such as 'hiraeth' – a term of longing for reconnection with this geographies, cultures, and stories of Wales, is definitive of Welsh culture. The concept is suitably explained by Elmes (2005:98):

> "'hiraeth' [he-rye-th] is that sense of belonging or homesickness that Welsh people feel in their hearts when a long way from the Valleys of the south or the mountains of the north. Now it may be that this special sense of emotional attachment to a place, of longing and belonging, is much the same the world over... and maybe this sense of magical connection to a country and culture is simply yet another of those myths of 'localness'. Yet I'd bet there's scarcely a single Welsh man or woman who doesn't recognise... 'hiraeth' as representing something for them that's very special and very, well... Welsh".

There is no direct equivalent of 'hiraeth' in English[4] although the notion of 'hearth' may come closest to capturing the fondess of home inherent in the term. However, 'hearth' does little to convey the broader sense of nostalgia, loss and belonging which is conjured up by hiraeth. As Pearson and Shanks go on to tell us, hiraeth is just one example amongst many that facilitates the connection of the Welsh to the landscape:

> "In Wales there are a number of spatial notions which describe the Welsh sense of place. These notions operate as a series of cognitive maps. *Y filltir sqwar* (the square mile), the intimate landscape of one's childhood, that patch of ground we know in a detail we will never know anywhere again.... *Cynefin* (habitat), that area where we feel we belong, the immediate environment, the surroundings which impress themselves upon us in the formative years between 5 and 15.... *Y fro* (neighbourhood, home district, heimat), where, as Thomas ... observes (1991:85), 'everyone in the community is joined to everyone else by a mesh of stories and incidents if not by family relationships'." (2001:139)

---

4 Although the German language does have 'heimat' whose closest translation may be to a sense of home place, whilst the Portuguese language has the concept of 'saudade', and Romanian the notion of 'dor'.

This joining of people, place and stories into entanglements of plot is identifiable in much Welsh culture. As author Grahame Davies tell us, "if two Welsh people meet each other outside Wales the first thing they do is ask each other where they come from, you know, specifically *which place* do you come from" (PC). In this way individuals seek to locate themselves and each other in this "small country... of coincidence and connection" (Clarke, 2008:9). The Welsh language therefore serves to emphasise and encourage the person-place connection, in Lopez and Gwartney's words, "it is a language that keeps us from slipping off into abstract space." (2006:XIII)

Janus Wales: Facing the future

> "'Fuckin' hell, butt' said Mazz... 'Glad to be Welsh then?'
>
> 'Was ist das?' said Lawrence in a dumb German accent and Mazz started laughing again.
>
> 'Christ' he said eventually, 'great to know you've got a culture that boils down to one famous play, two sixties cabaret stars, a male-voice choir and a twelve-year old opera singer, innit?'." (Williams, 2006:269)

So whilst the 'inward looking' face of Janus Wales looks to the past and perhaps a static sense of Welsh place, the outward face of Janus Wales looks to the future. As Cardiff writer Lloyd Robson states, "the world we live in is changing" (PC), and as a consequence questions are inevitably raised whether Wales's cultural past is fit for the future. Can relations between people and place be imagined beyond the *Milkwoods* (Thomas, [1954] 2003) and *Green Valleys* (Llewellyn, [1939] 2001) of the past? And if so, what might these relations look like? It is possible that the Wales of the future may be a fragmented, relational, and mobile place. In this scenario, how secure is the Welsh connection to landscape in the twenty-first century? Through looking to literature this book explores the new 'trajectories' offered by global culture and what relations between people and place are constructed as a consequence.

Walking plotlines

In order to enter into the complex entanglement of page and place, this book explores eleven contemporary Welsh authors writing in English who have composed contemporary narratives that explore the relations between people and place. Each chapter brings together three key sources: firstly, authored narratives which seek to take and make particular geographies through fiction; secondly, contemporary geographical theory which offers new perspectives on these narratives, and thirdly, interviews with authors to understand the importance of place in grounding and defining their stories. Through the combination of these three sources, each chapter investigates the relations between people and place; how characters engage with place to create a sense of belonging, where threats to their sense of place originate, and how they are engaged with. Each chapter is internally structured around a plotline – a geographical journey through the entangled geographies of fiction. Each plotline can be followed on the maps given in the text, or followed in practice through walking the plotline in the 'real' locations of Wales. Wherever possible the plotlines themselves were generated through collaboration with the authors interviewed, and walked with them in practice. Box 1.3 outlines the 'talking whilst walking' approach (see Anderson, 2004a) used to bring together the real and imagined in these instances.

Box 1.3. Walking the plot

"That's the way to see the world: in our own bodies." (Snyder, 1990:99)

According to Ingold (1993:152), walking can be considered a form of 'dwelling', a practice which immerses individuals within a locational context and reveals a range of lived worlds. Such an approach complements the aims of understanding 'plot' as it moves beyond the 'sterile opposition' of configuring geographical landscapes as neutral backdrops to human activities, or as cognitive or symbolic orderings of space. Rather it configures geographical landscapes as constituted by person-place relations, and as such prioritises the, "knowledge born of immediate experience, by privileging the understandings that people derive from their lived, everyday involvement in the world"

(Ingold, 1993:152). Walking the plot alongside the author therefore sought to produce not simply a conventional interrogative encounter, but a collage of collaboration: an unstructured dialogue where all actors (interviewer, author, and place) participated in a conversational and geographical pathway creation (see also Anderson, 2004a). As a consequence, the knowledge produced is importantly different: atmospheres, emotions, reflections and beliefs could be accessed with regard to the plot, as well as intellects, rationales and ideologies.[5] This book is therefore able to enter into this entangled landscape, engaging with, or calling forth, particular plotlines through practice. Through the method and means of walking, we as researchers, readers and writers can bring together the visible, visual and vicarious (after Riley, 1997) and walk into worlds where meanings are accessed and produced.

In *Page and Place* plotlines are suggested through the nation's writers and places. These plotlines are not definitive, simply illustrative, and you are encouraged to engage with them, walk them, and through doing so further accrete your own routes and roots to them. Through practice, these plotlines will inevitably produce more alternatives, inspiring "endlessly intersecting narratives [with] tale-tellers interrupting tale-tellers, [and] lives that fade into other lives" (Sinclair, cited in Pinder, 2001:15).

In detail, *Page and Place* is structured around four parts: 'Capital Cardiff', 'Kaleidoscopic Aberystwyth', 'Nations', and 'Borderlands'. Each seeks to explore different aspects of the ways in which Wales is responding to the 'responsibility' (after Bauman, 2002:XV) of "produc[ing], stag[ing] and cobbl[ing] together" (Beck, 1994:13) new relations between people and place in a globalised world. In 'Capital Cardiff' four authors are discussed: Peter Finch, Andrew Craig Williams, Lloyd Robson, and Tessa Hadley. The plotlines of each author focus on a different area of the capital (Cardiff Bay, central Cardiff, Splott, and Roath respectively), and each give a different insight into the ways in which people create connections to the places around us. Finch (Chapter 2) focuses on how contemporary

---

5 Such an approach resonates with the psychogeography tradition (for more, see Chapters 2 and 12).

places can be read as armatures of the past, how collective histories that have been silenced or marginalised can be called forth by imaginative archaeology to contextualise our place in the present day city. Williams (Chapter 3) echoes Finch's ability to imagine the lives of the past, but does so through focusing on the individual rather than the collective. Williams concentrates on personal artefacts and experiences and the meaning these give to our identities. His writing demonstrates how cities appear to develop personalities through our engagement with them, and through our own individual mixtures of friendships and experiences, we are able to create a familiar home in a new place. In turn, Robson (Chapter 4) focuses on the complex relations that individuals can have to particular (hyper)locales, and the tensions that generate both pride and frustration in our connections to places. He shows how a particular hyperlocal politics can be generated through spending time on the streets and, through forming ties in this way, places can be become as important as family to many people. In contrast, Hadley (Chapter 5) focuses our attention on the lack of belonging individuals can feel in places, particularly in suburbs. Her characters explore the possibilities for regenerating a sense of community – through focusing on the mundane and everyday, or escaping to 'liminal spaces' – and becoming renewed in a fresh set of person-place relations. By focusing on these plotlines we see how places soak up stories, and how over days, months, and generations, places become increasingly complex as they co-exist in the present. As Italo Calvino reflected in *Invisible Cities*, it is possible to identify these stories if one is sensitive to the cues held by the location:

> "A description of [a city] as it is today contain[s] all [a city's] past. The City, however, does not tell its past, but contains it like the lines of a hand, written in the corners of the streets, the gratings of the windows, the banisters of the steps, […] every segment marked in turn with scratches, indentations, scrolls." (1986:10)

In these coincident places our lives become not simply part of the here and now, but are also contextualised and given a new perspective when we consider they are also part of the there and then. Our connections to place are therefore not simply a product of the present, but also influenced by our sensitivity to the past. From these plotlines we can see how places can be understood as coincidences of events, memories and artefacts, remarkable for being simultaneous and connected across generations. Coincident places are, therefore, individual and collective, and it is in these reams of fabric, time and space that we weave our own lives.

In Part II, 'Kaleidoscopic Aberystwyth', we see how this university seaside town can be experienced through the plotlines of Niall Griffiths, Richard Collins, and Malcolm Pryce. Griffiths' Aberystwyth is a destination for the drug-addicted and the drifter, a home for those without hiraeth, but a place in which meaning and identity can be fashioned through a primal connection to wild nature (Chapter 6). For Collins (Chapter 7), Aberystwyth is an unnamed location in his novels, but nevertheless provides a rich topography through which lives can be remembered and retold. In this way, Collins explores ideas of truth and belonging, to both people and places, and the ways in which these connections can be refracted and changed through our memories. Malcolm Pryce (Chapter 8) adds a layer of parody to the place of Aberystwyth through his Louie Knight novels. Juxtaposing the American noir thriller to the sleepy seaside, Pryce brings a rich mixture of humour, affection, and mystery to the Ceredigion coast. These chapters illustrate how there is something in Aberystwyth's geography that invites the imagination and "lend[s] itself to endless reinvention" (after Griffiths, 2008). In this Part of *Page and Place* we see how despite offering the potential to generate belonging, modern cultural life can also alienate us. Nevertheless, belonging can be found in alternative person-place relations (be they in wild nature, the meeting of land and sea, or even a re-imagination of the past). This Part illustrates, therefore, how it is possible to generate belonging in novel and unorthodox ways, and how these processes offer hope to the ennui encountered in modern life.

In Part III, 'Nations', we turn to the poetry and prose of Gillian Clarke (Chapter 9) and Grahame Davies (Chapter 10). Both writers locate their work deep within local places across the length and breadth of the country, and do so in order to tackle issues of nationhood in contemporary society. Clarke and Davies engage with processes of mobility and globalisation and how these offer both threats and opportunities to belonging in Wales in the twenty-first century. Through focusing on her own life experiences, Clarke attempts to bring into being a new form of belonging at the national level. She reminds us of the importance of inhabiting a local place and becoming sensitive to its rhythms and history in order to value these processes in the face of 'progress'. Through a fictional story of a modern day campaigner, Davies extends this idea by showing how insular modes of belonging – however vital and nourishing – can be supplemented and strengthened by the opportunities offered by globalisation. Through these plotlines, we see how becoming a tolerant and cosmopolitan nation does not necessarily involve displacement and disorientation, but can allow the appreciation of ourselves and others whilst also strengthening the connection between people and place.

In Part IV, 'Borderlands', we focus on the poetry and plays of Owen Sheers, and the psychogeographical novels of Iain Sinclair. Both writers locate their work on the Borderlands, not only between England and Wales, but of the past and the future, the real and imagined, and between belonging and non belonging. Sheers (Chapter 11) writes from his personal experiences growing up on the borders. In his poetry we see how he found his place through connection to the rural lifescape, but – in his novel *Resistance* – we see how this connection can be undermined by forces from abroad. Sinclair's pyschogeography explores his own exile and estrangement from Wales (Chapter 12). He demonstrates how the processes of mobility and modern life can fracture identity and senses of belonging into often contradictory and incompatible shards. In this Part we see how the borderland identity can in some ways encapsulate the tensions and trials that are inherent within Welsh (and perhaps all) national identities in contemporary society. In attempting to be both outward and inward looking, we are able to see the merit of both sides of belonging – of being insular and static, and open and mobile. Our identity, in short, becomes protective and progressive, phobic to some change, whilst philic of others. It is in this tension, explicitly found on the border, that we find the nature of belonging in the twenty-first century. Though this literature, therefore, we can see how it is possible to forever strengthen and evolve a sense of self and geography.

By looking to literature to give us poetic and often oblique insights into the nature of belonging in the twenty-first century, we can see how we are attached to a variety of scales, yet it is the hyperlocal – be it a hill, a street, a café, or a pier – to which we attach our lives. But as we do so, the infinite plurality of ideas that imagine our connections to place come into play. Some ideas are based on accidents of birthright, or inherited blood ties; others based on new notions of indigeniety (through, for example, language, everyday practice, tax-paying, or chosen affiliation). These competing ideologies of place attachment can divide as well as connect, even within the same place. Thus by looking to literature to understand the relations between people and place we see how location and identity are formed through a complex entanglement of ideas and practices. We are geographical beings, and the places around us influence our sense of well-being, sense of identity, and ultimately our belonging. Through connections with people, stories, and practices, places form the stage and an active presence in our everyday, storing our memories and storying our lives. Taken together, our entanglement of belonging is open to new possibilities – refusing to be limited by old voices and outdated connections, and willing to go beyond them to open up plural opportunities for the future. Thus in a context of multiple scales and means of belonging, it is unlikely that person-place relations can ever be a singular and unified thread in this entanglement. Person-place relations, be they individual or collective, are not a 'finished object', and (like the literatures in this book) are always a 'dots' in a '*trail of action*' (Ingold, 2005). The geographies of fiction therefore offer a vital insight into this state of ongoing composition, offering us the opportunity to freshly assess these important relations in our own lives. By exploring page and place as an ongoing composition of plot, we can become "exposed to some of the connections and contradictions that are meaningful to our position" (after Williams, 2002: 184), and as Williams outlines in relation to her own family life, identify what it vital to us and our future:

> "In a room on the other side of the Atlantic a baby is born. She is a peach like Ma. Her eyes are Bethesda blue and her hair is cornfield yellow. My daughter names her Ruby and in moments I shift a whole generation. I shoulder the burden of memory as the generation of my father passes away and a new generation is born elsewhere. I can't write Ruby's history. Eventually she will pull together her own pieces and make sense of them in her present. I sit and wonder what exactly I must pass on to her. I try to piece together the parts of the collective memory, the essence of my generation that may be meaningful to hers. It is our quest to dig, to pull down the idea of return like a faulty gene; is it the sea stories and the stories of movement; is it the story of unbelonging? The story of global connections and yet local attachment, of boundary disputes, of changing the very idea of race? What it is that binds us? How will we reinvent this heritage together?" (Williams, 2002:190-1)

Part I

Capital Cardiff

Chapter 2
Stalking the soul of the city: a finchian plotline through Cardiff Bay

> "Mazz was riding the Ghost Train. The 125 from Paddington to Cardiff. Ghosts everywhere. Ghost behind him in London.... Ghosts ahead of him in Wales. His family, what was left of it, up in the valleys, his mam long gone and passed away, his dad a ghost for years, haunting every pub that'd have him from Newbridge to Newport. Ghosts in Cardiff." (Williams, 2006: 213)

Our first plot- starts at the end of the -line. The 125 from Paddington to Cardiff ends for Mazz, a character in John Williams' *Cardiff Dead* (2006), in Cardiff Bay. As Williams (2005) tells us, Cardiff Bay was once known as Tiger Bay, and this neighbourhood, alongside the adjacent Butetown district, are distinct areas separate from the city of Cardiff itself. As Williams suggests, these places have their own ghost narratives (see also Smith, 2011).

> "Tiger Bay was the original pirate town. The way I heard it as a child, the real old-time pirates, Captain Morgan and his crew, used this promontory off the then small town of Cardiff as a base: a little piece of Britain that was beyond the law. God knows whether it's true or whether my addled memory has simply cobbled together a new myth out of two or three old ones, but still, it's a legend that suits the wild side of the Welsh." (2005:217)

At times a 'fortress', a 'ghetto', an 'island', and an 'isthmus' (Finch, 2002), Cardiff Bay and Butetown have been racially, historically, and physically divided (by the river and the railway) from Cardiff itself. As a "world of its own" (after Defreitas, in Williams, 2004), the area has a rich and varied history which has often been ignored, silenced, or destroyed by the development of the city and the Bay. As Williams states:

> "So what that Tiger Bay is now remembered only as history – replaced by something new and shiny, and determinedly without history, called Cardiff Bay?
>
> Well so quite a lot – without memory and history we lack community and we lack soul. Community is something Cardiff has always been good at. But Tiger Bay gave [the city] its soul. So let us remember it for sure, but let us also try to preserve something of its anarchistic,

> tolerant, cosmopolitan soul in this new place that we have learned to call Cardiff Bay." (2005:234)

Both Mazz and the 'Cardiff Bard' (Finch et al., 2006) Peter Finch remember Tiger Bay. The area's past, its history, and perhaps even its authenticity, is for them wrapped in myth, story and imagination. As Mazz begins to uncover his past in *Cardiff Dead*, one story in a trilogy that is "set in a city called Cardiff that broadly resembles the real one" (Williams, 2006:vi), Peter Finch does so in his *Real Cardiff* series, three books that are set in the city and its environs. The *Real Cardiff* series offer an antidote to the gleaming surfaces of sanitised tourist brochures and, as we will see in this chapter, they seek to get under the skin of the city and root around for marginalised pasts, reawakening urban ghosts in order to better understand, and retain, the 'real' soul of the city.

The Millennium Centre to Bute Street

> "This was the point at which schizophrenia arrived. In Wales, we celebrate our schizophrenia. We have north/south schizophrenia, we have urban/rural schizophrenia.[1] We permanently seem unable to work out who we are." (Finch et al., 2006)

We start our plotline in the icon of the new Bay, and perhaps the new city: the Millennium Centre.[2] It is tempting to suggest that this is both temporal and geographical point at which the Janus faces of Wales come into contact (see Chapter 1). At the turn of the century, where the Senedd government building looks out to the Bay, and the Millennium Centre boasts the "the best five word poem in the English language" (Griffiths, 2008 123)[3]: "in these stones horizons sing", Cardiff Bay is at once an echo of the city's connection to its industrial past, as well as a symbol of its devolved future; not only does the Bay

---

1 And as we have seen in the Introduction, a past/future schizophrenia too.

2 As Finch recounts in Real Cardiff 2, the Millennium Centre is "Jonathan Adams' storm-defying piece of essentially-Welsh, landmark architecture. Is it like bend-wall Bilbao? High-rise New York? Glass-wave London? No. It's steel and glass and slate and ancient timber and, at the back, walls six-stories high of what looks like creosoted lap-wood fencing. It's Wales. Not England. What we have. How we are" (Finch, 2004:148).

3 This poem is written by Gwyneth Lewis.

signify the country's routes to the world, it also remembers its roots to the landscapes and resources of its hinterland. This is the meeting place of the past and the future, of the near and the far, of the fictional and the real, and of the many faces of Cardiff and Wales. It was at this 'point of schizophrenia' that I met Peter Finch, who had kindly agreed to take me on a tour of the 'Real' Cardiff Bay. Exploring the many faces of this meeting place has become, for Finch, an 'obsession' (Lee, 2010).

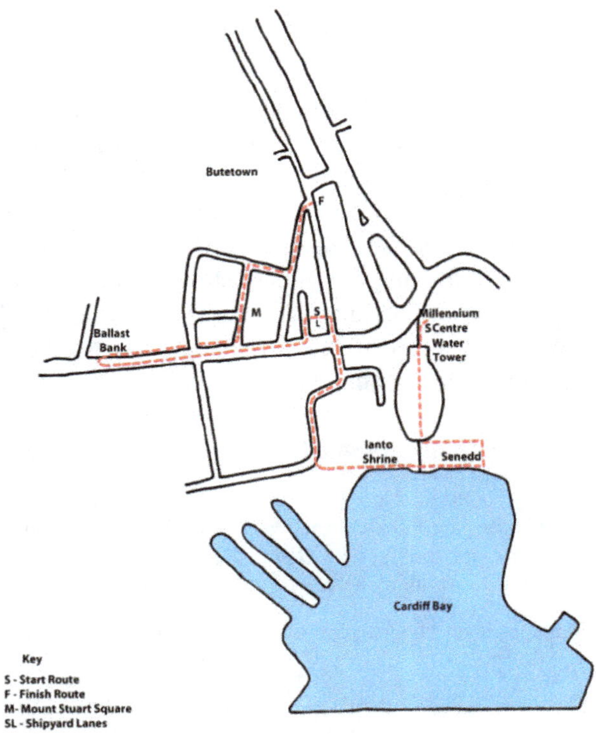

Map 2.1 Millennium Centre to Bute Street

Figure 2.1 Millennium Centre, and view from Senydd across the water of Cardiff Bay. (Source: Author)

Meeting Peter Finch at the Millennium Centre under the words *'Creu Gwir Fel Gwydr O Ffwrnais Awen'* I am reminded of the importance of location and language to the fabric of Welsh culture. As geographical beings, the question of *who we are* is intimately connected to *where we are* (see also Anderson, 2004c), and humans must understand their language, culture, stories and geographies in order to fully engage with the various faces of their identity. Through his writings, Peter Finch has become intimately aware of the necessity for this engagement. His poem 'We can say that' (Finch, 1987), begins to explore the tensions in Welsh identity with reference to the problems and opportunities presented by the Welsh language:

> "We speak the language. No, We understand it. We say that. The bricks of our houses are thick with it. We know all the songs and the place names. It's not foreign. We flaunt our origin in the big city where most don't know iaith y nefoedd from Urdu. We do.
>
> Heol y Frenhines it says and we can say that. We are native. We have status.
>
> Cenedl heb iaith cenedl heb galon until we have to handle it in Penrhyndeudraeth or Pwllheli where, using English slyly, we say we're glad it isn't dead." (Finch, 1987: 33)

In 'We can say that' Finch identifies how the Welsh language is integral to the fabric of Welsh culture. Cymraeg has become imbued in the stones which sing of horizons, and has been built into the foundations of the nation's homes and hearths. As a consequence its cadence is familiar, and there is something in saying 'Queens Street' (Heol y Frenhines) with the correct timbre that bestows indigenous status and difference. However, Finch also identifies how expertise in the language is not something shared by all, particularly in Cardiff, and English is often the language of choice for everyday communication. The remains of the Welsh language are thus valued, but they are a minority weave in the tapestry of the culture. Indeed, Finch goes further to suggest that any strong preoccupation with speaking Welsh is part of a version of the nation's identity that roots it to the past and suppresses the potential for new words and identities to be imagined. In another early poem 'A Welsh Wordscape' (1987), Finch confronts what he sees as this oppressive version of Welsh identity by parodying the form and style of RS Thomas' 'A Welsh Landscape' (see Evans, 2013) and thus challenging the limits of the nationalism debate:

> "To live in Wales,
>
> Is to be mumbled at by re-incarnations of Dylan Thomas in numerous diverse disguises.
>
> Is to be mown down by the same words at least six times a week.
>
> Is to be bored
>
> by Welsh visionaries
>
> with wild hair and grey suits.
>
> Is to be told
>
> of the incredible agony
>
> of an exile
>
> that can be at most
>
> a day's travel away.
>
> And the sheep, the sheep, the bloody flea-bitten Welsh sheep, chased over the same hills by a thousand poetic phrases all saying the same things.
>
> To live in Wales is to love sheep and to be afraid of dragons.
>
> …

> [It] is taking a long time to learn that yesterday cannot be today and that the world is fast becoming bored with language forever in the same tone of voice.
>
> Look at the Welsh landscape,
>
> look closely,
>
> new voices must rise,
>
> for Wales cannot endlessly remain
>
> chasing sheep into the twilight." (Finch, 1987:15-16)

Railing against the preoccupation with hiraeth and language, Finch identifies the need to look again at the (literary) landscape of Wales, and realise that – in the twenty-first century – its voices should not limited by the 'same words' or shadowed by the literary figures of the past. As a response to his own call, Peter Finch has become one of the 'new voices' that have arisen to reinvent the literary geographies, and perhaps even culture, of Wales. His works seek to make new connections and directions from the meeting places of geography and language, creating cartographies for a country and city that is changing. We discuss these new directions in relation to his series *Real Cardiff* on the first steps of our plotline from the Millennium Centre to Bute Street.

> *I ask Peter Finch about the origins of the Real Cardiff series?*
>
> "I was invited by the Editor of the Planet magazine (back in 1999 I think) to write something about the changes that have happened in Cardiff. Now this was interesting as I wasn't known for that sort of writing and I was quite surprised to think that if the changes that are happening in the city can be seen by somebody sitting in Aberystwyth then they must be worth looking at. So I went out to think about what I was going to write and I put together a piece about the development of the city, the changes that had happened in this place.... It took me back to places I haven't been to for a while, it made me think about personal history, it made me think about historic history ...and it is in this combination that the first *Real Cardiff* came from." (PC: Personal Communication)

With the *Real Cardiff* series Peter Finch seeks to explicitly engage with the changes occurring in the social and material form of the city, documenting where the city has been and exploring the directions it is taking in the present day. By engaging with the past and future in this way, the series examines the changing identity of the city. Finch has written of the *Central City* (2002), the *Greater City* (2004), and the *Changing City* (2009), and the success of these books has led to the series proliferating across Wales. Although they are titled 'Real...', true to the nature of a schizophrenic or Janus-faced city, the books combine a 'reality' which merges strict ontological fact with stories, myths, and personal histories. *Real Cardiff* is therefore an entangled mix of elements rather than a factual guidebook. For Finch this is an intentional and political statement. His "general distrust of guide books" which "rarely [give] you all the information you want" (Finch et al., 2006) directed his approach to the series. As he states,

> "I was a bookseller for many years... and my bookshop specialised in books about Wales … the thing that seemed to mark out the publications of local historians was the combination of two ends of the spectrum: one … would just present reproductions of lists from ledgers of the period without really locating this properly or interpreting it in anyway, or at the other end the sentimental tales of how we used to play 'rat tat ginger' when we were kids on the street. I got so bored of this. I thought 'this is too sentimental, too nostalgic and too badly written. There must be a way of writing about location and the people who lived in that location that isn't just nostalgia and is maybe entertainingly or engagingly done', and that's what I wanted to do.
>
> I wanted to uncover that thin layer of city, the place which exists underneath the fabric that we see around us now. The comparison I took was between Pembrokeshire and where we are in Cardiff. In Pembrokeshire the landscape is all still visible, it may be built-on but the ancient past and prehistoric standing stones all exist, you can go to them, they're marked on maps, the legends are alive in those landscapes, the myths that residents and others have built on top are all there. Come to Cardiff and it's all gone. For two reasons: one, we built on it, and two, even if Cardiff had not been built on and someone had put a stone circle here by now it would have sunk into the bog lands because what we are walking on is reclaimed territory. Why do you think the Romans built the castle where they did? It was because that was the best place to build; anything further south was bog land. So I walked around the city looking for bits that could make me think 'how did this use to be?' I wanted to find the stuff that was underneath, I wanted to find the myths, the ley lines, the sources of power, the places that actually felt like they were something magic, something ancient, something that were full of people that are no longer with us." (PC)

As Finch identifies, the very geology of the city – as an entity that floats (see below) – contributed in part to the gradual erosion and loss of its history. However, perhaps more important to the loss of the soul of the city was the regular razing of Cardiff's built form and concomitant erasing of the voices and identities of the past. Finch's attempt to uncover the historic soul of the city thus focused not only on the haunting remains of social ties but also on the remnants of the city's built form. For him both persist as a thin geological layer under the present day geography. Finch's attempts to uncover the soul of the city is perhaps best understood as a form of imaginative archaeology, a process which seeks to evoke the relations between the past and the present through materialities visible to naked eye, or through atmospheres or recollections which are only hinted at through physical cue or a sympathetic invention (see Box 2.1). It is also a process which places Finch within a broad 'psychogeographical' tradition. Psychogeography, as Pinder states, "involve[s] a variety of explorations of the physical and psychological landscape of the city" (2005:386). Bonnett (2009) recounts how 'psychogeography' derives from the Lettrist and Situationist Internationals who sought to reimagine everyday space. Walking, or 'purposeful drifting' around a landscape was an attempt to explore the physical and psychological aspects of a place (see Pinder, 2005). There has been a recent revival of the psychogeographical tradition in more traditional geography and literary circles (see for example, Sinclair, 2002, and see Chapter 12; Ho, 2006; Home, 1997; Keiller, 1994, 1997; see also Ackroyd, 2000, 2007; Moorcock, 1988), but in short the process arises,

> "as one of a set of ideas and practices ...of how places affect the psychological states of those who pass through them. With a reciprocal meaning: that the places might be changed in order to change the experiences and mental states of their residents and visitors. This was part of a theory of radical activism for the transformation of cities." (mythogeography.com)

Box 2.1: Imagining Tiger Bay.

Figure 2.2 Tiger Bay mural welcome. (Source: Author)

As Peter Finch and I walk through Tiger Bay he points out key material cues to its history:

> "This [Finch points down an alley in Cardiff Bay, see Figure 2.3] is an old shipyard lane and I think this is very reminiscent of what Cardiff must have been like, the way it bends and the shape of those buildings there, the curved tops on the door, if you try and think of this without the cars and without the dustbins and the ends of those air conditioning machines you can imagine it. So I did a lot of this, walking around the city looking for bits that could make me think 'how did this use to be?' and get me into that mind-set." (PC)

> "We are now on the bottom of Bute Street and why, for example, did this railway come down here? [see Figure 2.4]. There used to be engine works over there, why were they over there? Actually the first Bute Dock was over here, that was filled in and buildings have been put on top of it but that was the very first dock and I tried to envisage what it would have been like with [the second Marquis of] Bute sitting on his horse coming down the side of it. He couldn't see very well and had these thick glasses. I spent a day once standing over there trying to pretend I was Bute, I'm Bute on my horse now, what does it feel like?" (PC)

Figure 2.3 Tiger Bay's old shipyard lanes. (Source: Author)

Figure 2.4 Cardiff Docks and County Hall Council Buildings. (Source: Author)

> "The Avenue runs parallel to the site of the former Bute West Dock, the break for maritime fortune made by the second Marquis of Bute in around 1839. This is where Cardiff first began to stretch its economic muscle. ...King coal was arriving. Like a nineteenth-century Bill Gates the Marquis made his early move, established the operating system, and took over the city. Docks and the land they're dug into. Own both and you control the world." (Finch, 2002: 150)

Over this history, contemporary Butetown lives its days, entangling itself with the events of the past; as Finch writes:

Figure 2.5 Butetown murals. (Source: Author)

"Bute Street is [now] slowing down. The chicanes, narrows and parking bays of urban traffic calming have made their appearance along its length. The render on the Queen Street-Butetown rail link embankment retaining wall has been chipped back to reveal clean Victorian stone. Before the railways this was mostly tide-field meadow, marshland drainage ditches, fishing henges, the Cardiff south moors.... No one lived here much until Bute drained it for his great docklands. Calling it Tiger Bay came later. Portuguese seamen reputedly sailing though the rough waters of the Bristol Channel said reaching Cardiff was like sailing into a bay of Tigers. Did they? Who knows. The name rose, stuck, then, when they pulled the old streets down to replace them with high rise, fell. There's the Tiger Bay Cafe on the bright new frontage and they are planning a Tiger Bay Bar for the new Millennium Centre but the name is not much on the lips of the locals." (Finch, 2002: 148/150)

In contrast to Pembrokeshire where the local's "legends are alive in those landscapes", Peter Finch sought to revive Cardiff's myths and histories, or at least re-place them with some of his own. To paraphrase Coverley, Finch's 'Real…' series amounts to, "a complex [Cardiff] project which incorporates poems and writings that revive the lost histories of the city" (2006:121). In this way, Cardiff's geography has become,

> "reconstituted through a superimposition of local and literary history, autobiographical elements and poetic preoccupations, to create an idiosyncratic and highly personal vision of the city." (Coverley, 2006:121)

Although Finch did not fervently follow the Situationist politics of psychogeography, like Iain Sinclair he was keen to integrate the personal alongside the collective in his *Real Cardiff* series:

> "For me, the first person was very important.... I didn't want to sentimentalise anything and I didn't want to just recount history, I wasn't interested in that. But by the same token I was interested in why a place was significant and the sort of historical structure that caused it to be like it is." (PC)

Alongside this imaginative archaeology a 'literary eclecticism' is also evident in Finch's series, in both inspiration and practice. Influenced in part by Torrance (1980) and Fisher (2005), Finch was also inspired by a range of American 'Beat' authors; as he writes:

> "Kerouac consumed me. And his fellow travellers Ginsberg, Corso, Burroughs. My copy of *Protest,* the British anthology of Beat writing published in paperback by Panther, sanitised Ginsberg's earth-shattering, amazing and, for me, totally engaging *Howl.* I'd never come across anything like it. I went out hunting for the real version. Found it, unaccountably, on the shelves of the SPCK Bookshop in The Friary, Cardiff. Ferlinghetti's City Lights Books edition, San Francisco. Cord jeans. Railroads. Wide open space. Trails going somewhere. For the heart and for the spirit. The Kerouac scenario. I could do this. But couldn't. This was Wales where the roads went up valleys and got lost among pits and sheep and trees. Far less space. Much more rain." (Finch et al., 2006)

Figure 2.6 Finch's San Francisco beat entanglements: City Lights Bookshop and Kerouac traces on the City's streets. (Source: Author)

> "The battleground was not how to engage with the wider universe (or even with the inner one) but with the greenness of our landscape and with its clattering tongues. Sod this. I had no history or reputation to get mashed. I could do what I wanted." (Finch et al., 2006)

Inspired by the radical inventions in style and structure of the Beat poets, Finch's own poetry and eventually his 'Real…' series adopted a multiplicity of voices, interweaving 'historic ['factual'] history' (see above) with, "some of my own, inevitably literary, autobiography" (Finch et al., 2006). In this way Finch indigenised the 'Beat' style for his own city:

> "America is rich with people's histories and the[ir] histories of places seem to be riddled with literary anecdotes around references to the art rather than the local trades or industries. We don't have that for Wales, and we certainly don't have anything like that for Cardiff. Cardiff seems from the histories and the books I've read to be almost a blank sheet as though no writers have ever come from here and no artists have ever considered the city as anything worth recording in any way and I thought I'd rather like to correct that." (PC) (see Box 2.2)

Box 2.2 You are here: Mount Stuart Square

Inspired in part by these US authors, the *Real Cardiff* series began engaging with place in different ways. We arrive in Mount Stuart Square, and as we do so I can hear words from Finch's imaginative archaeology in my mind: this is a 'city that floats'.

Figure 2.7 Mount Stuart Square: Are you here? (Source: Author)

"I'm in Mount Stuart Square, solid Victorian, once home of shipping companies, coal traders and merchant banks. Over the past one hundred and fifty years, as Cardiff slowly moved from industrial powerhouse to postmodern mall, the Square and its occupants have changed. Gone is nineteenth-century commerce. Gone too twentieth-century grit, dabble and decline. In their place are twenty-first century liberal arts - broadcasters, blond-floored apartments, bars, media companies. At their centre the monolithic Coal Exchange, once vibrant with top-hatted traders selling options in not-yet-cut Aberdare steam coal, and now home to concerts by Captain Paranoid and the Delusions, Jefferson Starship, Van Morrison and Otis Grand."[1] (Finch, 2003:13)

Figure 2.8 Coal Exchange, Mount Stuart Square (under renovation). (Source: Author)

"...Today, in the Ship and Pilot, Darren, or someone with a name like that, tells me he's not out to waste his useful time talking to muppets. He knows I'm not from round here. This is a small community. Everyone knows everyone else. I'm from Lisvane. No. Well, you're from th'apaartmunts, then, you should be drinkin in Bar38. Norere. This aint the Bay. It is not. I buy him a drink, no use falling out over nothing, dangerous in deep Butetown, whatever the night. He has a double Southern Comfort and Coke. I don't take my notepad out. The Bay is two communities existing in the same space-time continuum. They pass through each other as they move. They rarely interact." (Finch, 2003: 15)

> "As a child this part of the Docks always frightened me. Huge water discovered suddenly in the heart of the built-up city, or so it seemed. I'm trying now to put it in place, make something from it, fit it into the internal urban landscape I'm creating. Turning what's in front of your eyes into something that can be written down isn't easy. Do what Joyce did. Put the whole day down as one. Interior, exterior, first thoughts, best thoughts, all of them crammed onto the valid page? Go with Glyn Jones' contention that, if you let it, the one idea you have will just roll like a snowball, gathering other things to itself, growing rapidly until it's either topsy or done? Or be like considered Sebald or demented Sinclair, walking with your notepad, letting your mind go off in whatever direction it chooses? ...Somewhere in the contemporary surface around me, or just under it where I can still reach if I stretch, is the real city. A shimmer of the past, still there in the ether, just, in the whorls of dust, in the way the earth has settled, in the dried surface marks and in the etchings of rust. Catch it, touch it, hold it. Be quick before the mind moves. Fickle. On." (Finch, 2002:17)

In line with the tradition of psychogeography, Finch creates his eclectic mix of real and imagined 'plotlines' on foot. As ('demented') Iain Sinclair states, "Walking is the best way to explore and exploit the city; the changes, shifts, [the] breaks in the cloud helmet, [the] movement of light on water" (1997:4). Like Sinclair, Finch is wedded to the physiological engagement with place that walking enables, as he states:

> "I discovered that walking is the best way [to explore the city] because as we are now, you walk, you are touching the territory, you can put your hands in it, you've got a physical contact. It's not just the speed you are walking at, it's the fact that you are part of the landscape, the landscape makes you tired, it makes you cold. You are also pretty visible when you are walking and you get challenged on occasions... but you can write anything down and you are not in the cocoon of your car".
>
> *I ask Finch whether walking also affects the style of his writing, its rhythm or pace?*
>
> "Yes, yes it affects its pace absolutely because if you are on a train for example your writing is totally different. The train's got a rhythm to it, it makes a noise and that somehow gets into what you are writing, whereas contrast that with walking, it's very, very different. You have to give yourself time, and this way you find things." (PC)

The rhythm of walking, discovering, and writing combine in one of the key 'found things' we came across on our walk around Cardiff Bay: Ballast Bank.

> "I'm bringing you here to show you this. This is my poem." (Finch, PC)

Figure 2.9 Ballast Bank. (Source: Author)

"...The history of the city that I write is part of a post-modern process which results in poems like this. The current text we are standing in front of represents a specific part of the history of Cardiff: this is what remains of the Ballast Bank. The Ballast Bank was made when early wooden ships came into Cardiff in order to take out iron and coal for the docks when all we had was the canal basin. They'd come in with the ballast in the hull of the ships, they'd take the ballast out and they'd load up with iron or coal and off they'd go. Little wooden ships we're talking about here, you know under one thousand tonnes, and they'd throw the ballast on the bank of the basin and eventually there was a bank of this ballast something like three quarters of a mile long. There's an old map that exists dating from the very early 1800's which shows the ballast bank and when the docks came to be developed where we are now, the ballast from the bank was used as foundations and structure for some of the buildings. There is a church in Charles Street for example and if you look at it it's got all sorts of different stones, it's made of piles of ballast stones." (PC) [see Figure 2.10]

Figure 2.10 Church in Charles Street, Cardiff, reputedly constructed with stones from the Ballast Bank, Cardiff Bay. (Source: Author)

Figure 2.11 The constituents of Butetown: Ballast Bank and Butetown Murals. (Source: Author)

> *I ask Peter Finch whether it's really the case that earth from all across the globe can literally be found in the foundations of this city?*
>
> "That's exactly it. So I said: why don't we put the ballast bank back? This is stylised [of course], but the poem is made by words derived from the races, languages and activities of the people that flowed in and out of Cardiff in the past 200 years. The Portuguese petroleum came in, the Poles came in, the potato trade came in; there were Indians, there were Irish, there were Italians, there were people from the Forest of Dean came and worked here, the Danes worked here, the Hispanics and so on. There are also a couple of jokes in this. The word Vulcan is in there, which related to the vulcanising process, there were some vulcanising works over here, [but] it was thought I meant the Vulcan pub (for more, see below), so that word had to go.
>
> *I ask him whether those who commissioned the statue didn't see it as a reference to intergalactic 'Star Trek'?*
>
> "No! The word Vulcan was censored from the piece; however, they missed this one. Klingon." (PC)

Finch's imaginative archaeology thus not only scratched beneath the surface of the city to discover the history of Cardiff Bay, but also attempted to reinsert this history through memorials such as The Ballast Bank. As a consequence of the historic and contemporary references called forth through engagement with the 'Bank', as we walk away from the memorial I begin to think of The Vulcan pub, and the different communities that once lived in the neighbourhoods of Tiger Bay. As we walk through the contemporary streets I begin imagining the historic phases of the city, and with every turn of a corner I fold over geographies and histories to see the [now dismantled and relocated] Vulcan Pub, and Newtown, the forgotten Irish enclave of the city.

Figure 2.12 There's Klingon's on the Ballast Bank, Captain. (Source: Author)

Box 2.3. From Ballast Bank to The Vulcan

"That's what Cardiff does, it knocks everything down." (Robson, Personal Communication, see Chapter 4)

Figure 2.13 The Vulcan Pub, Newtown, Cardiff. (Source: Author)

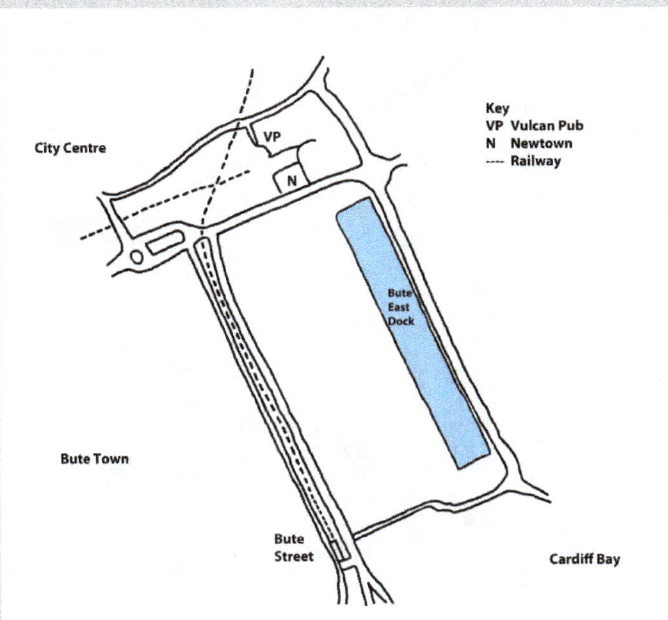

Map 2.2 The site of the Vulcan Public House

As we have seen, Finch attempts to investigate the soul of the city, scratching at its surface to see how the new and the old piece together the nature of the city in the present day. Although keen to make sure the city has room for 'new voices', Finch has developed a strain of nostalgia for those elements of the city that are removed from its landscape. One key landmark that is mourned is The Vulcan pub. When we undertook our plotline of the city, The Vulcan remained; however, within months – and despite a long standing campaign to save it – (see http://save-the-vulcan.blogspot.co.uk/) it was dismantled in 2012.

> "I get upset when I see things that I think of as important being just knocked down because somebody wants to build something else. I do recognise that you cannot keep everything, you have to have progress, but we seem to allow money to drive what happens rather than social need or any kind of personal interest in something.

> There's no concern or respect for maintaining culture in any way. We're by the Vulcan pub now and I don't know whether you know about the campaigns that were fought to save this place? It's still here, it was about to be knocked down about four years ago, a great victory, but it's under threat again now. I can't see it staying, it will eventually go; there are plans for this site to be turned into apartments. That'll happen, just not yet." (Finch, PC)

Figure 2.14 The car park which has now been expanded onto the site of The Vulcan. (Source: Author)

Fellow Cardiff writer Lloyd Robson (see Chapter 4) shares Finch's disquiet at the ease with which the city wipes clean the key material forms that define its heritage:

> "Well Cardiff kind of knocks down the quality stuff and replaces it with kind of cheap shit and then kind of goes on about what a great historical city it is. There's so many fantastic buildings and sites that have been knocked down for the sake of the St. David's 2 [shopping centre] for example. Cardiff is its own worst enemy and it wants to knock down The Vulcan on the way into town. One of the great Cardiff pubs and they will probably replace it with a car park and then kind of build down the street a themed Vulcan, you know? It will knock it down and then glory in the nostalgia of it. That's what Cardiff does" (Robson, PC).

Interestingly, in May 2012 the BBC reported that, following its removal, The Vulcan was going to be rebuilt at the Museum of Welsh History in St. Fagans, Cardiff (BBC, 2012).

Box 2.4. From Cardiff Bay to Newtown, Tyndall Avenue

"Can a whole district – houses, people, even the name – simply disappear?" (Finch, 2002:165).

Figure 2.15 Writing over Newtown. (Source: Author)

"Ask your average Cardiffian if they know where Newtown is and they'll tell you it's in mid-Wales.... [Yet] scrape the Cardiff surface and you'll find shamrocks underneath." (Finch, 2002: 165)

"By the time Newtown was marked for demolition in 1966 it had 169 falling-down houses, two pubs and a garage. Half its population called themselves Welsh rather than Irish, but their names O'Sullivan, O'Leary, Burns, O'Shanahan, Dwyer – gave away their origins. ...Families were dispersed to Ely Pentrebane, Trowbridge and Llanrumney. The community was broken." (Finch, 2002: 168)

"The district we're in now by the way is what was Newtown. It was the Irish district of Cardiff. It was all close communities, tiny streets, back to back houses, about a dozen pubs in this small area which extended from the railway line there to the round here the top end of the dock. It was all pulled down in 1960s. Where you see the crane over there to the flyover at this end, a tiny place, very Irish. You see the word potato in the poem of mine [on Ballast Bank] down the bottom. Just over here is a converted warehouse, this has been converted into apartments called England's. The potato warehouse is where the potato trade worked. The barges would come up the West Dock, the first dock, load it with potatoes which would be offloaded there, put into sacks and the Irish people who lived in this area would work there." (Finch, PC)

I ask Peter Finch if the workers lived here because of that trade?

"They lived here because of that trade and the docks themselves. They originally came here to navigate the canals, some of them to dig the canals, then to dig the docks, then to work in the docks, and then to work in the ancillary trades, you know, shovelling coal. These were what were called in historical terms 'walking suburbs' because in those days nobody had a car, there was no public transport. If you didn't live near where you worked, you didn't work. So they tried to live as near as possible to where they worked so the original walking suburbs of Cardiff were Grangetown and Newtown, where people could walk from their houses directly to the docks that employed them. So that's why the Irish lived here and of course a lot of the population of Cardiff has Irish origins and it's in part where the accent comes from." (PC)

"[This area] is called Atlantic Wharf on the Agents' plans. They're selling apartments at a new property they're calling *City Wharf*, top of the dock, eastern corner, on a footprint about the size of a corner shop. Onto this they've managed to cantilever eight stories. The name is vital. No *Caerdydd* anything. No mention of Dock either. Far too redolent of work and smoke and coal and pain." (Finch, 2004:153)

Figure 2.16 Atlantic Wharf this way. (Source: Author)

"All that's left today are the fourteen standard-pattern council houses built late on the bobtail, waste ground once fronting Tyndall Street. This was a red-light district for a time after the prostitutes were moved up off the newly cleansed Bute Street. They are west of here today, near the Central Station. I ask someone mowing their front lawn if she'd heard about little Ireland. Yes, she had. Mrs Kitchen next door was here when the houses still stood. So what's the area called today? What do you put down when you write your address? Atlantic Wharf, she replies without a flicker. The mower roars." (Finch, 2002:168)

Imaginative archaeology, radicalism and nostalgia

> "Cardiff is forever looking for itself. For the places of its origin. For the spots where history launched. For the locus of just what made it. …It all flows through here." (Finch, 2009:136/8)

The imaginative archaeology which re-places and remembers community landmarks like the Ballast Bank and Newtown could be understood as simple exercises in 'radical nostalgia' (see Bonnett, 2009). Radical nostalgia, according to Bonnett, is prevalent in psychogeography as writers bemoan the loss of history within a city, arguing "against the superficial trend towards modernity" (Heartfield, 2004). For these psychogeographers, "the [utopian city] is, above all, a backward-looking one. The[y] adore everything arcane and archaic about the city" (ibid.). Despite Finch admitting to getting 'very excited' (PC) when he finds undiscovered remains through his imaginative archaeology, his 'Real…' series is more than a reactive 'Cardiffnostalgia' (to paraphrase Heartfield, 2004). In line with his early challenge to the inward, limiting tendencies of Welsh culture, his work acts a bridge between the Janus faces of the city, linking the Bay to its marginalised origins in order to position its contemporary identity. His documentation of Newtown (Box 2.4) is a case in point, reminding the twenty-first century city that it is placed on the waterfront, and those places beyond the horizon have been, and continue to remain, vital to its vibrancy and life. As Finch states, this imaginative archaeology reminds us that:

> "In some senses Cardiff is made up entirely of people who are not from Cardiff, it's just the lengths of time that are different. I mean Cardiff was less than 2,000 people two hundred years ago so all the others have come from somewhere else. Those 2,000 people didn't just procreate, it didn't work like that, everybody has come from somewhere else. ...I wanted to reflect the fact that this place was not as Welsh as it would like to make out [or that it is defined by different form of Welshness, see also Chapter 3 and 10]." (PC)

In scratching beneath the surface of the built form of Cardiff, Finch is providing a link to neglected or forgotten pasts that crucially define the nature of the city, and by extension perhaps the country, today. As we have seen in relation to the Ballast Bank, the foundations of Cardiff literally have the soil of the world in them; the city is built on its connections to the world.

Box 2.5. Plotlines and faultlines: Cardiff Bay and the space time continuum

The times and places that Finch uncovers in order to find the soul of the city are not only those places and times of the past, but also those of alternative futures. This is when our Cardiff Bay plotline crosses a space-time faultline.

> "Almost everyone knows that there's a fault in the space time continuum which runs right through Cardiff.... The fault is the Cardiff Rift, a wormhole in space and time. ...The past is seen through it. The present gets sucked back. Fragments of the future appear and disappear like they were ghosts." (Finch, 2009: 122/3)

The point of our meeting, outside the Millennium Centre in Cardiff Bay, is the site of the Cardiff Rift. As Finch's Klingons colonise the Ballast Bank, so other (science) fictions cohabit the Bay. Torchwood, a BBC spin-off from Doctor Who, is sited in, and under, the Water Tower in Roald Dahl Plaza (see Figure 2.17). There is even a memorial to one of the series' fictional characters in the Bay (see Ianto's Shrine, 2013).

Figure 2.17 The Water Tower, Cardiff Bay. A portal to new worlds? (Source: Author)

As Finch might say, this is the "spot where history is launched… it all flows through here" (Finch 2009:136/8). This fault in Finch's plotline emphasise how plural fictions, realities, geographies and histories become entangled into the soul and substance of Cardiff Bay.

Conclusion

Following a plotline through Peter Finch's *Real Cardiff* outlines how histories, mythologies and story-telling help us locate our sense of place in the modern world. As forces of economic and cultural reorganisation provoke change in the capital, Finch seeks out the spectral and lets them give voice to the histories and stories that have formed Cardiff Bay's past and present. As John Williams has reminded us, "without memory and history we lack community and we lack soul" (2005:234), and Finch broadens our sense of community by including the past and hidden histories into our present context. Through doing so he threads stories and mythologies into the entanglement of plot; to paraphrase Bonnett, he "explores and re-imagines the forgotten nooks and crannies of ordinary landscapes [in order] to re-enchant and re-mythologize prosaic geographies" (Bonnett, 2009:46).

Walking the city and calling forth forgotten pasts reminds us how a city absorbs and stores both personal and collective memories. As Calvino tells us:

> "the city soaks up [our lives] like a sponge and expands. A description of [a city] as it is today should contain all [of it's] pasts." (Calvino, 1986:10)

In this reading, cities are affective, living, landscapes. As stones sing of horizons, houses stores language, and ballast becomes part of the city's material soul, a city's identity can be called forth by documenting lost pasts into our contemporary plotlines. In this way, Finch seeks to memorialise these transient pasts. As the past slips into the metaphorical and literal bog on which Cardiff is built, Finch seeks to create memorials to forgotten episodes; both materially in terms of the Ballast Bank, but also in terms of his written works. Finch's 'Real…' series can thus be seen as documentary armatures which allow the past to be seen alongside the present, co-existing but not necessary limiting the future. Through so doing, *Real Cardiff* also emphasises how the city is continually changing, as Finch states, walking the city:

> "proved to me again and again that in any of this psychogeographic rambling, in any of this process of walking through a landscape, you can't think you've done it because the reality is always changing. That's happened to me loads of times and I've discovered that what I'm walking through is very different from how I imagined it or remembered it." (PC)

In this way, Finchian plotlines are not simply exercises in nostalgia with one eye to the past, but are a constant reminder of Cardiff's Janus faces by having two eyes to the future. As the city changes around us, we note how 'Cardiffnostalgia' is never far away from 'Cardiffhope' for the future, as demonstrated by Mazz's girlfriend Tyra on his return to the city in John Williams' *Cardiff Dead*:

> "Wales are about to play Western Samoa in the Millennium Stadium which has somehow opened on time and forever changed the way Tyra saw the city she'd spent her life in.
>
> The old stadium, the Arms Park, [was still there, but] was almost invisible from the centre of town. ...The new stadium dominated the centre and you could see it from almost anywhere in Cardiff. And, like pretty much everyone else, Tyra found it surprisingly inspiring. It suddenly made you aware that Cardiff was changing, that all the bollocks you heard people spit on the TV about being a European capital for the new century was really true. Even made you believe that all the bay development could come to something." (Williams, 2006:443)

Finch's *Real Cardiff* thus draws our attention not only to the invisible histories of the past, but also to the new voices that will determine the city's future. His writing is not the definitive city, but by calling forth a city that is absent from official guides, Finch channels the beat spirit and has produced a 'Do it Yourself' (or D.I.Y.) version of the city (see McKay, 1998). As Finch's voice memorialises the life, times, and soul of Cardiff Bay, other new voices contribute their own plotlines to the neighbourhoods and suburbs that constitute the city itself. It is to these voices that the book now turns.

Chapter 3
Plotting relations: writing roots into the heart of Cardiff

> "Each living human being... is for himself [sic] the centre of the universe, with lines radiating out from this unique centre.... These lines are his relationships with the world around him. The most important can be plotted." (Gwyn Williams, 1977:16)

As Peter Finch walks through 'Newtown', away from Cardiff Bay's soul and into the city, our plotlines lead into the heart of Cardiff. This central district and its immediate suburbs are where many people feel the pulse of the city. Modern day migrants, from Wales and beyond, find their place here and begin rooting their relations with the world around them. Whilst Finch was concerned with mapping the histories of the city's soul (see Box 3.1), others seek to centre their present life in this new urban universe; one is writer Andrew Craig Williams.

Box 3.1. Urban histories 'versus' personal plotlines: Peter Finch 'versus' Andrew Craig Williams

I ask Peter Finch whether it's true that there should be canals on the streets of Cardiff?

"Yes there should be, and sometimes I'm driven by an obsession to find their remains. There's the dock feeder for example which runs through Bute Park. We're just coming into the city centre now and I wasn't aware until I did the research that the Bute feeder is still here underneath Churchill Way. It was built to put water into the West Dock and to flush out the sediment and I got obsessed by this, tracking it through the city. If you walk along Queen Street there's a little lane called Park Lane which if you walk up about half way up on the left hand side there's a gap in somebody's car park and if you peer over the wall you look straight down there's the water still flowing. When I discovered that it was an exciting moment. You know I found another trace of the past that's still here." (Peter Finch, Personal Communication (PC))

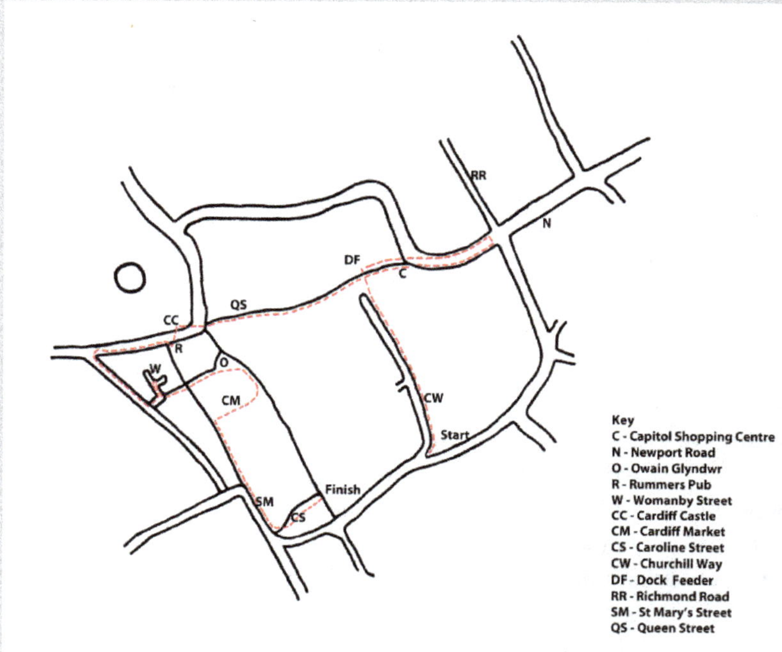

Map 3.1 Cardiff Central

Figure 3.1 Finch's hidden Dock Feeder. (Source: Author)

"I live in Splott, on a tiny street in a tiny block near Moorland Park [see Figure 3.3]. When I had my offer accepted, I turned into a Time Team detective; spent hours on Ancestry.com and old-maps.co.uk. I discovered that in 1890 a lady called Ellen Rörstrum lived in my house, and was probably the first occupant. When my Dad and I removed the old suspended floor from under the stairs, we found a rusty old Victorian hatpin. Part of Ellen's life was suddenly in my hands. I felt I knew her. I could see this woman bringing up the children who survived past infanthood, mourning the ones who didn't." (Andrew Craig Williams, 2011)

Andrew Craig Williams' plotlines thread and fray with the urban histories of Peter Finch. In the short story *Rosie's Last Week* (2005), Williams' character Geraint literally falls over into Finch's plotline as he trips over a "wonky overflow channel" [see Box 3.2] that flows into Finch's hidden Dock Feeder. I was first introduced to Andrew Craig Williams' work in the *Big Book of Cardif'* (2005). When I initially moved to the city, trying to find my own place in its heart, I was taken by the key role that the city played in the plot of his short story *Rosie's Last Week*. At the time, I did not know the places referred to in this story (beyond the landmarks of Cardiff Castle and The National Museum, see Box 3.3), but I was drawn to the idea that Williams' characters had built up a relationship with the city, and that Cardiff appeared to care about them in return (as we will see below). Now, as a resident of ten years, I am familiar with William's streets: his plotlines have become part of my city. I've walked them, stumbled into them, and taken tours of them (see Box 3.4.); and it was in the Rummers pub, a key location in *Rosie's Last Week*, where I interviewed Williams about his city plotlines (see Figure 3.3).

Box 3.2. Rosie's Last Week

"IT'S RAINING TIGHT LITTLE LINES on me, as I'm skipping over puddles on Richmond Road. My flares get soaked through every time I'm unsuccessful in my leaping, water and brown leaves sticking fast to my shoes.

My toes get wetter, but I keep on jumping, I have to meet Rosie before eight or she'll kill me. She told me she's got something special planned for tonight but she wouldn't say what. I turn quickly on to Newport Road, into the honk and blare of cars ripping through the wet tarmac. I pause under the viaduct to light a fag, and two girls walk past me with a Gucci umbrella held between them,

"You look like you're bleedin' love!" says one of them. I look down at my red t-shirt, and the large drops of water make the red even darker like I've just escaped a psycho killer in a slasher movie. I smile as they pass by giggling and flicking their hair like manes; one brown, one pink." (Williams, 2005:202)

Figure 3.2 The honk and blare of Newport Road, and the clock, Queens Street, central Cardiff. (Source: Author)

"Rosie told me to meet her at the clock on Queen Street, just like we used to back when the food hall was still open in Capitol. Every Wednesday afternoon would see us sipping coffee at The Continental clutching Poppy Z or Phillip Larkin, eyeliner and attitude, dark, dyed hair and bright smiles. I'm running now, trying to get there before her. Scruffy pigeons fling themselves out of my way casting accusatory glares at me with their reptilian eyes, suddenly I can smell hotdogs and the whiff of engine and for a second, I'm back at the fair.

I can see Rosie ahead of me, she has no coat either and she's wearing her wine-coloured velour top, the one she knows I like because it makes her look sexy. She knows because I make her wear it so often. I try a last burst of speed to reach her and fall headlong into a rivulet caused by the wonky overflow channel in the slabs on Queen Street. I look up into her concerned face, pale skin glowing in the half-light, one eye tracing a mascara line down her cheek, and smile." (Ibid.: 202)

Figure 3.3 The Rummers Pub, central Cardiff. (Source: Author)

"I'm expecting to get drunk tonight, hopefully sit by a radiator, and talk shit with my best friend. We stop outside Rummers and Rosie hugs me,

"You'll be dry soon." She says and we step inside, I'm dreading the stares and cheers but manage to keep my composure as we make our way to the nearest radiator. I cling my arms around it grateful for the warmth as Rosie goes to the bar to get our drinks. I re-arrange myself so that my back is touching the rad, my eyes close in happiness." (Ibid. :203)

Central to Williams' writings are themes of connection, memory, past, and place. Where Peter Finch focuses on how material objects like buildings, and their decaying remnants, can offer us portals to the past, Williams focuses on more everyday items – like hatpins in the case of his own house, or smells in the case of hotdogs and engines – and how these can store the memories of our and others' lives (see also Clarke, Chapter 9).[1] If Finch is interested in how 'stones sing', Williams

1 In doing so, Williams echoes Gillian Clarke, in her poem 'Confiant'(see Clarke, 2003:54, and Chapter 9), with Ellen Rostrum haunting his hearth in the same way as Clarke herself does in her imagination in the old family house.

focuses on how the 'everyday evokes'. Yet for both writers material things can be agential and affective, stor(y)ing the lives of people and offering us a way into connecting to the past and bringing it into the present. As Williams writes in *Rosie's Last Week*:

> "'Do you know Tarina?'" Heledd asks me, and I explain that I work with her boyfriend Kahu. We talk about Heledd's job in the University's Archaeology department, and she promises to take me on a guided tour. I tell her how excited I am at the prospect of this, she laughs at how eager I am. ... I tell her how amazing it would be to hold an actual arrowhead from Mesolithic Wales.
>
> 'We don't have many human remains though,' she tells me, 'but plenty of animal ones'
>
> 'Oh my God!' I exclaim a bit too loudly, I am starting to feel slightly drunk, 'So you mean I could actually hold a sheep's tooth or something that actually grazed at an Iron Age farm?'
>
> 'No problem.' she tells me, seeming genuinely pleased that I am so chuffed to bits. I offer to get the next round in and I stand up, Heledd moves her skirt out of my way and her knuckles touch Rosie's hand, which is reaching into her bag. I see the two of them exchange a quick peep at each other. Rosie keeps looking, but Heledd breaks eye contact first. I see this millisecond of action take minutes." (Williams, 2005: 204-5)

Here Williams highlights how time can be concertinaed in the most everyday of conversations. He illustrates how material objects like hatpins or arrowheads can pierce through time and space, and how simple tactile gestures can slow time as a lived experience. These material objects and affective practices are thus part of the relational ties (as Gwyn Williams notes above) that form our 'universe'. These ties are like 'strings' (after Calvino, 1986:76) which stretch back through time and across distance, tying us into generations and geographies. Even when the owners of those strings pass on, the ties remain and echo the existence of those relations. As Blair et al argue, "memory relies on [these] material and/or symbolic supports" (2010:6); it is through these material objects that recollections of another time and place can be evoked (after Belk, 1990: 670). Through this act of memory, or to be more precise this practice of imaginative archaeology (see Chapter 2), those lives and materialities can "regain some measure of [the relations'] living substance" (Dorst, 2013:36).

Material objects can thus prompt imaginative archaeology, but places can too. The land can hold our memories, and offer them back to us, in the same way as material items. The Irish poet Seamus Heaney was sensitive to this possibility, writing that:

> "I began to get an idea of bog as the memory of the landscape, or as a landscape that remembered everything that happened in and to it", and words, or indeed objects, could act as "doors... a point of entry into the buried life of ...feelings." (Heaney, 1980:54/52)

Place and objects within them can therefore story past lives. As Gwyn Williams (1977) put it specifically in relation to Wales: *The Land Remembers*. Calvino suggests the landscape of the city in particular is "like an armature" (1986:15), storing our memories as well as holding histories that we can access if we observe and imagine in an appropriate way. As a consequence, place becomes not only a process of ongoing composition, but also an entangled 'coincidence' of events, emotions, memories and artefacts remarkable for being simultaneous and connected (see Anderson & Moles, 2007). Integrated into the armatures of place are individual and collective experiences that are at once contemporary and historical as well as real and imagined, and through these armatures people ground themselves, find meanings, and tell the stories of their lives. 'Coincident places' are not necessarily a 'compression' of time and space (Dicken, 2003), or a 'folding' of geography and history (Doel, 1999), but more a stitching together of the reams of fabric of time and space into which we weave our own lives. Williams' imaginative archaeology therefore helps to access these coincidences, his writing draws our attention to the contemporary and historical lives of those who have "dwelt within [the city] and in so doing, have left there something of themselves" (Ingold, 1993:152).

Despite Calvino suggesting the city is "like an armature" (1986:15), Andrew Craig Williams suggests that the city does more than function as a cellular "aid to memory" (after Calvino, 1986:15). For Williams, the city is not only a passive receptacle but is always an agent, it has power and affect. Whilst places, landscapes, and localities may sediment personal and collective human meaning, and thus become key prompts for bringing back memories, Williams considers Cardiff to do more than this. For him, Cardiff is an actor – just like a human being, it is a place with a heart and personality.

The city as civilian, the city as subject

For Andrew Craig Williams' the city has personality. Due to the absorption of generations of relations and ties, the urban fabric has become redolent with lives, and even perhaps has a life of its own. As he states:

> "To me it makes sense to [talk about cities as alive] because for me a place has personality, just like a human. Cardiff is like this big hive mind. I've written quite a lot about Cardiff [as a character], you can't seem to help it if you live here. Most of my main characters live in Cardiff for a while ...either they are famous and well known, or (in)famous to me personally, and are [themselves] a facet of the City's character; every person, every shopper, every landlord, waitress or singer is a thought that flits through the City's mind. I have the same relationship to Cardiff as I do with the people I love." (Andrew Craig Williams, PC)

Akin to Williams, the geographer Yi Fu Tuan wrote widely on how places have personalities. He suggests that places, "like human beings, acquire unique signatures in the course of time. ...Loosely speaking, the personality of place is a composite of natural endowment (the physique of the land) and the modifications wrought by successive generations of human beings" (1996:445). Williams implies that as an active agent, the city can actually *produce* a type of human – every person, he says "is a thought that flits through the City's mind". I would suggest this falls short of claiming that the city conjures up humans (although one cannot discount a science fiction-style magician pulling all our strings through the Cardiff Rift, see Chapter 2), nor does it suggest that the city is the determining (rather than simply an influencing) factor on human personality (see Semple, 1911; Semple and Ratzel, 1941). It does, however, point towards the co-ingredient nature of people-place relations, and the poetically imagined interrelations that exist between geographies and identities. As Williams goes on:

> "I like to write a place as a living entity. London for example is for me a huge monster which swallows my friends, it swallowed my partner in there for a while, so for me it's a huge black monster which kind of swallows the people I love. Every place is different, and every place will have a different personality. I live in Splott... that's maybe not a very nice aspect of [central] Cardiff [he says smiling], but then you get places like Pontcanna which has got Pontcanna Fields and it's all

> lovely and you have, you know, art, boutiques, and Chapter [Arts Centre], that's kind of *The Guardian* reader part of the 'Cardiff Brain', and then you've got Roath which is the student area [etc, see Chapter 4 & 5]." (PC)

Cardiff's different neighbourhoods are therefore wrought from the hive mind by those who live in them (or, depending on how boldly you take Williams' analogy, the hive mind produces certain types of people). Whilst Pontcanna becomes a left-of-centre bohemian enclave, Splott is the central working-class heartland. Through residing in these places, individuals get to know and mould their own personalities. Places gain meaning through the social relationships in which they are involved (after Miller, 1998:131), and people in turn become defined by these person-place relations. Williams' identification of the personality of Cardiff and its neighbourhoods is perhaps a consequence of being an incomer to the city. As Peter Finch remarked in Chapter 2, 'in some senses Cardiff is made up entirely of people who are not from Cardiff' (PC), and the cosmopolitan nature of the city is something Williams comically notes in a passage from *Rosie's Last Week*:

Map 3.2 The streets and spaces of Splott

Figure 3.4 Moorland Park, Splott, and its surrounding streets. (Source: Author)

> "Let's go and eat it in the castle grounds," I suggest, so we make our way to the market to buy ham and tomato cobs. I imagine I'm a vampire looking at the various people in the market, wondering how they would taste. So many flavours for a Cardiff vampire to choose from: Welsh of course, Indian, Chinese, so many East Europeans, redheads, Aryan blondes, dark little Celts, Somali, Sikh, Swansea tastes. A boy couldn't make his mind up which to choose first. I would probably go for a redhead, pale skin and blue/green eyes, golden hair dusted onto icing white skin. I suppress a shudder and pay for my cob and diet coke." (Williams, 2005:208)

All the different nationalities which make up the Cardiff universe need to find their place in the 'Hive Mind' of the city. Williams' characters, Geraint, Rosie and Kahu all moved to the city, and through spending time in its heart, got to know its personality:

> "Why did you move to Cardiff?" she asks me, plucking some fragrant green leaves from a little baggie, the smell reminds me of rosemary.
>
> "What do you mean?" I ask.
>
> "Well, why Cardiff? Why not Swansea?" she asks me. I shrug my shoulders,

> "It's just the place you move to, isn't it?" I say. The sky is so blue today, and the noises of the city have a happy cheerful tinge to them, even the exhaust fumes take on a pretty violet colour. "Of course, I moved to be nearer you too, honey."
>
> "I was *waitin'* to hear that!" said Rosie, raising her eyebrows in a mock warning.
>
> "Why did *you* move here then?" I ask her as she lights up, the pale blue smoke makes me think of incense.
>
> "Well, after Uni it was logical to stay here. I just had to keep remindin' you to come and stay with me." We stayed silent for a while, each thinking about nothing much in particular. "I was so fed up of Llanelli, the whole small-town thing. Cardiff is the Metropolis of Wales, like." Rosie beamed as if she'd made an excellent discovery.
>
> "It's hardly a Metropolis though, is it?" I asked her, sunshine bleaching everything around us into washed-out colours. I recall Kahu saying something lovely the first night I met him about what he first thought of Cardiff when they moved here from New Zealand: he said it's big enough to be a city, but small enough to feel like a town. I didn't know what he meant at the time, but I get it now. I'm a smalltown boy like Kahu, and London scares me when I go there, but Cardiff has that town thing where old ladies who live next door to you call round with a parcel that they took off the postman for you that morning. I really love that…." (Williams, 2005:209)

By spending time in a neighbourhood of a city, Geraint, Rosie and Kahu physically encountered Cardiff and its personality. As we have seen in Chapter 1, these encounters 'take place', they affect and alter the nature of the location being visited, perhaps through changing or reinforcing its use, leaving pollution (or in the case of Figure 3.4 leaving chalk traces on the very streets), or simply through their presence and subsequent absence. This is a process that the phenomenologist Ed Casey has titled 'outgoing': a place becoming affected and formed through people's everyday activities (2001). According to Casey a reciprocal process also occurs, where places affect the people in them. This 'incoming' process involves the traits, customs and personality of a city becoming inscribed onto the human self (Casey, 2001: 688). As a consequence of these two processes, the personality of the place becomes part of the individual – as Casey puts it, "places [come to] possess us – in perception, as in memory… insinuating themselves into our lives" (Casey 2000:199). In his own life, Williams sums up this everyday process of incoming and outgoing and the way in which it generates his affection for the city:

> "I think the difference in Cardiff is that because it's so small you see the same people all the time, so it's almost like having the same thoughts all the time and you become kind of comfortable here.... You almost stop noticing the things as they become so familiar. But you put down roots through simple things like walking the same way to work and saying hello to the same people, or diverting on a Friday to the pub and back, and then on the Saturday it's walking to the park. The city becomes yours then." (PC)

In most cases, the processes of 'outgoing' (walking the same routes to work or diverting like clockwork on a Friday to the pub) and 'incoming' (as cultures of the city mould a person's identity) remain a metaphorical and affective insinuation. However on some occasions the process of incoming may *literally* occur, perhaps as Williams' characters take in the blue of the sky, hear the noises of the city, or breathe in the city's exhaust fumes. Such literal incoming is also implied in the following extract from *Rosie's Last Week*:

> "'You're not sleepin' are you?' asks Rosie and I open my eyes with a start.
>
> 'Just gettin' warm.' I assure her, and then see the girl standing with her. She looks shyly down at me on my seat and smiles, her eyes widen then close as she moves her gaze away from me.
>
> 'This is Heledd. Heledd, Geraint. Geraint, Heledd' Rosie introduces us.
>
> 'Hi Heledd,' I say shaking her hand, a few drops [of rain] fall from my arm and onto the small table. I notice one drip into my pint, imagining Cardiff pollution slowly spreading into the beer." (Williams, 2005:204)

Given the incoming and outgoing processes that the writer (and his characters) has/ve for the city, this reader imagines how this 'Cardiff pollution' isn't necessarily an unwelcome addition to Geraint's beer. Here the city may become something that is ingested and become physically part of the character's body, part of what the poet Adrienne Rich calls 'the geography closest in' (1986). In this way, the city for which Geraint has affection becomes part of his bloodstream and part of what makes him feel alive, as if he were a vampire gaining sustenance from its energy. Here then the process of outgoing and incoming of the city to the subject is complete.

As outlined in the above discussion, the personality of place is not simply something to acknowledge intellectually, it is something that generates emotion and affection. According to Tuan, in this way places become the geographical equivalent of a favourite old raincoat, comfortable, cosy and reassuring to engage with (after Tuan, 1996:46). Tuan emphasises the affective capacity of these reassuring and familiar relations between people and place through the neologism *topophilia* (Tuan, 1974). Topophilia refers to the affective bond – or relational sensibility as I have argued elsewhere (2009) – that is registered in a person but created by their co-constitution with a particular place. A positive relation between people and place (for example in a home setting) would be termed topophilia, whilst a negative feeling (perhaps in a strange or unwelcoming place) would be its counterpoint – topophobia.

Box 3.3. Topophilia and scale: national or local affection?

"It's nearly two o'clock now, and I'm waiting for Rosie outside a phone box as she calls her Dad. I can hear them having a mini argument so I walk on a bit, motioning to Rosie that I will be over there. I'm looking over at the castle walls, trying to imagine laying siege to it, how difficult it would be to conquer. I read in one of my Mam's history books about all the different ways invaders would try to get over castle walls, I'm thinking none would work on *these* walls: too high, too thick and probably too well guarded. I've never been inside Cardiff castle. I've lived here for nearly six months and I've never been inside. I was in a pub right opposite it last night and I didn't even look over at it." (Williams, 2005:207/8)

Figure 3.5 View of Cardiff Castle, from Rosie's phone box, and from Cooper's Field. (Source: Author)

As outlined above, although Cardiff Castle dominates the centre of the city, Geraint has never been inside. Through all his ingoing and outgoing in the city, he has never crossed this national landmark's walls. For him, this is a place for tourists or visitors, not for those who live and work in the city. This situation emphasises Williams' interest in the way geographical connections are dominated by the everyday and local, rather than the memorialised city (that preoccupies Finch, for example). For Williams, topophilia is generated in relation to more mundane landmarks in the city (like pubs, clocks, or even smells and experiences), and disinterest, perhaps even topophobia, shown towards these more notable landmarks.

Box 3.4. Womanby

As we have seen, Geraint and Rosie's attachment to the city is a product of their everyday behaviours and experiences. As young people, it is perhaps unsurprising that much of these behaviours are not only every*day*, but also every*night*, and focus on particular night-time hotspots of Cardiff. *Rosie's Last Week*, for example, begins in the evening, and is focused in part on Caroline Street (or Chippy Alley as it is colloquially known), St Mary's Street, and Womanby Street. The personality of Cardiff's night-time plotlines are well articulated in the following passages, the first by John Williams'*Cardiff Dead* (2006), and then by Peter Finch (2004).

> "Mazz had ended up back at the Colonel's. He'd gone into town after leaving Tyra, still pretty much walking on air. He couldn't believe what it was like once he got there. Friday nights in town he remembered there was hardly a place open, just a couple of night clubs and the casino, and the streets were deserted by midnight. Now it looked like a Hieronymous Bosch made over by Tommy Hilfiger. There were gangs of pissed-up boys and girls all over the place: gangs queuing up to get in and out of the dozen or so theme pubs and clubs in St Mary's Street, gangs roaming Caroline Street in search of dangerous looking food – that much hadn't changed at least – and gangs just throwing up in every alley and doorway." (John Williams, 2006:415/6)

Figure 3.6 St Mary's Street (by day). By night, 'Hieronymous Bosch made over by Tommy Hilfiger'. (Source: Author)

> "I'm entering the street of Saint Mary from the north. … Fairy lights on the lamp posts. Revellers in half dress. Bare midriffs, skin shoulders, bow ties, glitz and glitter. On match days the police cordon off the street to traffic and the roadway fills with rugby's pissed and surging thousands. It's like that now. …

"Ibiza uncovered in the south Wales gloom.... As I move south the sense of moral abandonment increases. Dinner-suited youths piss in doorways. A bare-chested shave head sprawls across the bonnet of a parked Mondeo. A middle-aged white-collar wearing a single earring and his shirt half out of his trousers falls off a traffic island. A girl in a short, sequined dress lies slumped again the window of a sandwich shop." (Finch, 2004:50)

"Crossing the junction with Wood Street, where buses bend towards the station and the edge of Riverside hoves into view, is like reaching the frontier. Surging hoards spin around the entrances to Wetherspoons. Black-coated, earphone encrusted bouncers watch from their doorways, overwhelmed. Fast-food boxes from McDonalds mix with cans and broken glass in the gutters. Someone comes out of the bowels of The Square, blinking, face full of blood, jacket smeared and ripped. A rubble of glass litters the floor of the philharmonic. A line of shrieking girls with lace and lycra and breasts that pop bear WKD with them as they stagger among the stuttering traffic. Coffee Republic has its shutters up, A fifty-year old with the velvet-collared fawn coat of a lawyer or wide boy lies on the pavement outside La Brasserie. Arriving diners, tanked, middle-aged, loud, cigar smoking, ignore him. Women of a certain age with flashing- Santa earrings and decoration tinsel wound into their hair wahoo in a conga line.

Figure 3.7 Caroline Street (left) versus Womanby Street (right) (by day). (Source: Author)

> Their chubby knees and fleshy shoulders shake in rhythm. An overweight on heels with five plastic carriers full of presents, a French loaf and a box of Sainsbury's crackers leans against the end of the bus shelter, a full large glass of house red in one hand, king size in the other. Music roars out of Walkabout. They are dealing again back of the Philharmonic. Guns'n'Roses. But it's Christmas. They won't go off. The world is going to end soon. We are all going to be there when it happens…. I consider retreating to Dorothy's Fish Bar in Caroline Street, last resort for all failed diners, but remember the rolling panic occurring on the pavement outside the former Cambrian. Give up." (Finch, 2004: 51/53)

Womanby Street is another key night-time haunt in Cardiff. Traditionally a less mainstream area, Womanby Street currently houses the fashion district of the city, as well as the Welsh club and more independent hostelries. Finch sums up Womanby Street in the following way, whilst Andrew Craig Williams contrasts the key differences between this area and the 'last resort' of Caroline Street in his unpublished poem:

> "Coming up Womanby Street in an August afternoon with the sun on my back. There are three women in front of me wearing heels that haven't been this high since 1961. They've got glittery bags big enough to hold a cigarette lighter and a lipstick. One's got celtic interlacing tattooed around her bicep and the other two have ankle chains. I've no idea if they are going out or returning home. This is the oldest part of the city, the dirtiest, the most exciting." (Finch, 2002:94)

Caroline Street vs.	Womanby Street
Chips	Burger van
Urine	Rain
Rats	Dogs
Fights	Queues
Darts	Cheap pints
Bottleneck	Back alley
Pedestrian	Guest-list
Cardiff slang	Welsh lang
White Concrete	Old paving
Wales shirts	One out, one in
Vomit	Screaming taxis
Falling	Over
Objective One	Mediaeval
Time to go home	Let us in
Middle Eastern	West Walian
New bars	Secret bondage club
L.E.D.	Watching the castle clock
Broken Windows	Bins
Workmen	Young men
There's a place you can go	I said, young men
Hayes	Dark days
Sunny overhead	Sunshine rays
Short	Long
New old	Old old
Awake	Tired
Anglo-Saxon	Viking hordes
Brothels	Pimps
Pixies	Imps
Finished	Abandoned

Andrew Craig Williams, unpublished.

Neighbourhood and nation: signs and simulation

As we have seen, Williams' writing focuses on the scale of the everyday, of the body, the neighbourhood and the (night-time) city. But Williams also imagines how these everyday experiences are scaled up across the nation. In the short story, *The Days of our Lives* (2001), Williams focuses on the mundane relationships and interactions between people and places that play out over the course of one evening and, in their simplicity and rootedness, define a version of the nation's geography.

> **"9:45 pm, Penybanc, Ammanford.**
>
> The Beatles are singing 'Nowhere Man', telling me not to worry, not to hurry, and that the world is at my command. Tonight, I'm inclined to believe them...
>
> **9:45 pm, Llantwit Street, Cardiff.**
>
> Emma's walking home on her own again. The tiniest buzz of a fading alcohol hit is finally swirling out of her brain, leaving her slightly thirsty. She snags her tights on an unkempt privet bush as her heels push her over another uneven paving slab,
>
> "Aw fuck!" she exclaims, throwing her rucksack onto the wet floor. Her house keys fall out of the small pvc bag as does her lighter, which proceeds to skitter across the slippery pavement and hurl itself into the nearest drain. She raises her eyes heavenwards and purses her lips. "Fuck it all."
>
> ...
>
> **9:59 pm, M4, just past the exit for Margam.**
>
> Sean is on his way to Newport. He's had a couple of days off work, and spent the time in Bridgend visiting his parents, drinking in the pubs he used to frequent as a youngster. The bleep of the heart monitor attached to his Auntie Annie's shrunken body still echoes through his mind, setting off miniature alarms of panic, of sorrow, and of terrible loneliness.
>
> All Sean wants to see is Teresa's black amphetamine eyes. All he wants to do is stop waking at four am expecting the phone to ring through bad news, or to find Teresa awake and smoking lights. Instead he thinks of his briefcase, and makes a mental note to remember to have one of the latches fixed before work on Monday.

10:04 pm, under Llandudno pier.

Stephen and Mark embrace in the cold, October air. Brief thoughts of school are blanketed by the smell of each other, by the weight of Stephen's waist in Mark's hands, the texture of Mark's lips against Stephen's. They are sixteen and seventeen (Mark is the elder, which entitles him to a locker and a prefect badge), and very cold.

....

10:47 pm, Llanstephan Castle, Carmarthen.

Dominic and Jay are smoking joints in the embrace of the castle's keep. Paint splattered hands cup the glowing red ends of their cheap soapbar joints. They've been here since about nine o'clock, making plans for total universal domination.

...

1:05 am, Escape nightclub, Northhampton Lane, Swansea.

Rhys is dancing with Erica. Big Al drops an old Jubbly's tune into the mix and the crowd goes wild. Erica holds her hands in the air and smiles at Rhys. Their shiny faces glow at each other, and Rhys realises just how much in love he is.

Erica doesn't think she can stand it anymore, and then the feelings grow more intense. They hug suddenly and she has tears in her eyes.

1:59 am, Penybanc, Ammanford.

I'm still here, in my nowhere land, writing all my nowhere songs for nobody." (Andrew Craig Williams, 2001:79-80)

Complementing Clarke's 'lanes' that 'lead everywhere' (see Chapter 9), in *The Days of our Lives* Williams offers brief portals into the everyday lives of people across the nation on an 'ordinary' Friday night. The melancholy of the Beatles' song 'Nowhere Man' infuses the piece, telling the narrator that the world is 'at his command', hinting too of the power of the author to control his character's destinies (contra Sinclair, see Chapter 12). Williams may feel like a nowhere man, but the places evoked are not no-where, they are to the reader now-here; their geographical locations root them in a reality which the reader can enter, however briefly. I tell Williams that I love the idea of a writer in a room on his own, perhaps a few drinks too many, wondering 'What are other people doing right now?' And then GoogleEarth-like, being able to see these lived moments, in real places, across the country. This is happening right now. This is our map of Wales.

> "I like that too. Who was that philosopher, Baudrillard I think it was, he had this idea that you can make a map so real that the story of every geographical point touches the real world. He warned that if we focus too much on the map, the real world just fades away. The real world becomes nowhere, and the map becomes real." (PC)

It is the significance of Williams' writing that he reminds us of the importance of staying in touch with the 'real world', with the practices of 'nowhere' people, which through mundane activities happening everyday in real neighbourhoods, mould our geographies and lives. If we remain sensitised to these relations, then our maps and plotlines will not displace or distort the 'real world', or become a sanitised simulation of them, but become continually informed by them.

In warning us of the perils of simulation, Baudrillard suggested that 'even the signs must burn' (1981: 163). In the case that concerns us here, it can be argued that these signs signify the relations that come to locate our universe. In William's writing, we have seen how the personality traits and signs of Cardiff are something he has an affection for, reminding us that the personality of places is not simply something to acknowledge intellectually but also something that generates emotion, familiarity, and comfort. In *Rosie's Last Week* this 'topophilic' affection is channelled into specific materialities which mark the neighbourhoods of the city that Williams' characters have grown to love:

> "On the way back to Roath, we stop at every street corner to look at the street names on the corners of buildings. It's a special font that I suspect was designed for Cardiff County Council, back when they were all put up, shiny and new and forever. I wish someone would turn that font into a downloadable one on the Internet, like they did with the London Underground one – I have it on my computer, it's called Paddington. If I had the Cardiff font, I would use it exclusively: for all my letters, in work, on my website, or even to write my shopping lists." (2005:210)

Figure 3.8 The 'Cardiff' streetname font. (Source: Author)

Through creating personal plotlines in the city, individuals no longer remain alien but the city becomes part of their universe and they become the city. Everyday 'signs', that are often noteworthy at first, become invisible as they become familiar, and they come to point the way to our belonging in place. These signs penetrate our lives, becoming reassuring waymarkers for our memories and roots. This interest and pride in particular street signs resonantly echoes a concluding section from John Williams' *Cardiff Dead*:

> "Tyra walked in and for a moment she thought there was nothing there. Then she looked more carefully and gasped in surprise. There, nailed to the walls, were maybe fifty, maybe a hundred, signs: street signs, a couple of pub signs, a few shop signs. Some of the street names she knew well; Maria Street, Angelina Street, West Bute Street, Sophia Street. Others, though, barely tugged at the edges of her memory: Nelson Street, Gladstone Street, Canal Parade. The pub names too – the Freemasons Arms, the Marchioness of Bute – were just memories. She knew what they all were, though. Her birthright. Her treasure. And evidence too. Evidence of what had once been here and was now gone, past, lost. ...she was sure [though] that in this end there was a new beginning." (John Williams, 2006:452/3)

As outlined by John Williams, it is often the case that material entities, objects and signs, as well as the memories and emotions they evoke, remain durable long after the places and people they once connected to have moved on. Although Andrew Craig Williams would champion the value of these everyday signs there is also an acknowledgement that, inevitably, these signs must too be destroyed. In order for the next generation of 'Ellen Rörstrums' to find their place in the city, contemporary signs must eventually burn.

This was precisely the case for Geraint in *Rosie's Last Week.* Geraint's topophilic roots, generated through outgoing and incoming to Cardiff with his best friend Rosie, were 'burned' at the end of her 'last week' in the city before her relocation to Australia. When she had gone, we see Geraint's familiar rootedness in the city, entangled with his relations to their favourite haunts, become displaced. The absence of Rosie shakes his sense of place in the city; the city becomes strange and alien to him – his sense of topophilia has changed to one of topophobia.

> "I've been without Rosie for nearly a week and it's killing me, no exaggeration, I really mean it. I haven't been this serious in God knows how long, she's taken my silliness away to Australia with her. I hope those Aussies are enjoying it.
>
> I'm sitting in the Owain Glyndwr, on the same table Rosie and me used to sit at. Even the table looks less clean now that she's gone, all of Cardiff has suddenly become messier and dirtier, nobody feels that they need to make an effort now that she isn't here to look at them or their tables. I've had one email and no phone call yet, I can't take being here alone so I get up to leave, maybe get a video out tonight and a bottle of wine." (Andrew Craig Williams, 2005:210)

Here we see how Geraint's identity isn't an insular and discrete entity, rather it is formed and moulded by his relations to people and to place. With the absence of Rosie, he is simply no longer the same person ('his silliness has gone'), and due to the Rosie's connection with Cardiff, it too seems a 'messier and dirtier' place in her absence. The impression is given that Cardiff as a place also understands its loss and it is no longer the same place without her presence. However, in a Rosie-less Cardiff, the 'hive mind' of the city steps in to give Geraint a sign to show that even though Rosie has gone, he is not alone.

> "The Slug and Lettuce looks busy, brightly coloured people spilling out onto the street, wearing crisp-packet clothes and pints of frothy beer on their heads, it makes me laugh. By the time I get to the top of Queen Street a clown has followed me and walks behind me pulling the same fed-up frown as I'm wearing. He sticks the tips of his white-gloved hands into the corners of his down turned mouth and brings out a magically produced paper flower, which he presents with a flourish. I accept it, and the smile that it brings to my face, gratefully. Outside Boots, a tall skinny young man plays guitar and sings, his guitar playing is amazing but his voice is terrible. I throw a fifty pence piece into his already quite full guitar bag, and he works the words 'thank

> you' into the Charlatans song he's singing. I smile at him and raise my hand.
>
> What are you doing? I ask Cardiff, in my head obviously, I don't want people thinking I'm touched. Are you trying to make me smile because you know I'm feeling sad? I'm at Sainsbury's now and there's a tiny man selling the Big Issue. I give him two pounds and tell him to keep the change, it is the guy's last one after all and nobody else was in a hurry to buy it from him. He asks me if I'm sure and I tell him yes of course, I'm happy that such a little, easy gesture can make a person smile like he is now. I sigh, and I'm still sighing past the Mayor's house on Richmond Road, she'll be back in a year.
>
> Don't worry Rosie, I'll keep Cardiff warm for you." (Williams, 2005:210/211)

Conclusion

In this chapter we have seen how Andrew Craig Williams introduces us to another plotline of Cardiff. Whilst offering us a similar 'imaginative archaeology' to the writings of Peter Finch, he nevertheless provides a counterpoint to the *Real Cardiff* series. In Chapter 2 we have seen how Finch excavates the city to discover its hidden and lost histories, whilst simultaneously depositing new layers through memorials in both stone and print. The *Real Cardiff* series thus redresses the balance of history disrupted by the straw man of the sanitised guidebook, and in the process reconstructs and refictionalises the city. Finch sets stones and words to mark, challenge, and conserve the city's pasts, creating memorials in order to (rein)force memory. In contrast, Williams wants to recall, celebrate and humanise the youthful city. Like the films of Gideon Koppel (2008), his writing is a "meditation on … belonging and a celebration of [friendships, socialising, and creating connections in place]" (Sleeve Notes, Koppel, 2008). For Williams, it is in the mundane that we find the meaning for our lives; in order to belong we must become anchored and rooted through a topophilic sense of place (after Ehrlich, 1985:5). Finch and Williams therefore approach the connections between the city and its people from different directions. Whilst Finch approaches from the side of the city, Williams approaches from the side of its people. In particular through *Rosie's Last Week* we see that the relations between people and place are crucial to belonging. We are reminded that everyday lives do not exist in a vacuum but happen in places, and these can be (re)lived as the city absorbs our stories. The personal becomes written into the physical, and through this process meaning is sedimented into this ongoing co-constitution. Despite the inevitability of geography in our lives, these writers remind us that there is also an invisibility of geography too. Finch warns how we often only notice our connection to place when it is threatened or lost: only when the signs burn, literally or metaphorically, do we feel our displacement. In turn, Williams reminds us we must not take place for granted; for him, as well as for Gwyn Williams, it is vital to plot those lines which radiate out from every human being and bind us to the geographies that define our place in the universe.

Chapter 4
Edge(y) territories: the hyperlocal world of Lloyd Robson

"you kairdiff? where ju live?" (Robson, 2001:41)

So far in *Page and Place* we have seen how the themes of memory, place, nostalgia, and change have been central to the entanglements of plot from Cardiff Bay to the central city. We now enter a different but connected territory, from the city centre along Newport Road (see Map 4.1), a line which is plotted by Lloyd Robson in his book *Cardiff Cut*:

> "newport road:
>
> gotta get home/try upping me speed, first i gotta cross this six laned street
>
> (rear red glow/undersill indigo/headlites shine lilac; futuristic vehicles hover eastbound carriageway, leave tracers of transit...)
>
> central reservation
>
> (shadows of infirmary; whiteclouds buzz smudge of blurred
>
> buildings in fronta me...)
>
> kerb & pavement." (Robson, 2001:56)

Travelling down this 'six laned street' takes us into Robson Territory, the hyperlocal triangle formed by the meeting of three neighbourhoods: Roath, Adamsdown and Splott. This is the 'edge territory' of Cardiff (1995); the edgy territory. Located on the periphery of the city centre but overlooked by many guidebooks, this territory sits on the edge of change and on the edge of chaos (sees Gleick, 1990). Robson's territory is not the 'heart' or 'soul' of Cardiff; it is the 'stain on the map' (Robson, 2003).

Figure 4.1 Crossing Newport Road, Cardiff. (Source: Author)

> "so this is my home, this morose dump this castle ground this moggy swamp this swab of land this marsh growth this tamponing town of urban blood & brown this festering hole this rising damp i keep running from & sloping back & insist on calling my home whether it wants me or not so this is what it's all about." (Robson, 2003: 95)

In this chapter we see how Lloyd Robson uncovers the role of place in forming his own identity. We will see how geography plays a familial role for Robson in forming his sense of roots and belonging, and how a political loyalty to the local and hyperlocal is mobilised through this process. Through an abridged plotline based on Robson's *Cardiff Cut*, the chapter examines the political importance of walking to engage with the city, and how this mode of attachment can generate tensions between notions of belonging at the local and national scale.

"'yu nose ow splott sounds bad ri? so we'r told any road, cos it' sounds liek SPOT or SPLATT!'" (Finch, 2003:21)

I meet Lloyd Robson in the Royal Oak, Broadway, half way along Newport Road. Peter Finch describes this location in the following way:

> "The main highway east, when I get to it, is Newport Road. This link passes on through the site of the Roath Court manor house's

gatehouse; past Cardiff's best unreconstructed Brains pub, The Royal Oak, with its second-floor boxing gym, its rock music back room and its heavy-booted regulars, and out onto what was once the causeway...." (Finch, 2003:22)

Map 4.1 The Edge Territory

As we settle down for a drink, I ask Lloyd Robson if this is his 'local' pub?

> "It is yes, in an awful lot of ways. It's a pub that I've either avoided or headed for all depending on certain stages and times of life but its certainly one that's been important to me even when I was a little kid because even when I was a child and I lived my childhood outside of Cardiff on the corner here they used to have a large blue neon sign advertising Brains [beer] and I knew whenever we came into Cardiff as soon as I saw that sign I was home. This is my territory from here on in so yes it's an important place to me – and a great pub." (Lloyd Robson, Personal Communication (PC))

Figure 4.2 The Royal Oak, Broadway. (Source: Author)

Robson's search for roots is intricably linked to this territory of the city. As we will see, this area of the city gave him a voice, a sense of community, even functioned as a family, and all these aspects of his plotline are made manifest in his writing, not least through his method of articulation. Robson's roots and depiction of place are articulated in his own vernacular, the dialect of the Splott / Adamsdown / Roath area of the city. His insider's voice is literally so, as fellow writer Niall Griffiths (see Chapter 6) sums up:

> "I'm not aware of anybody that does the Cardiff dialect like him; he gets it well. And he does a lot with it you know, you have to have a really quick wit, it carries some amazing images and visions… these dialects are repositories of knowledge so I think the more we keep these dialects alive, the more we keep a little kind of safes of certain types of local knowledge alive...." (PC)

In echoing the accent, the dialect, and the cadences of speech in his writing, Robson gives voice to an area of the city that is often silenced or overlooked; to paraphrase Spivak (1990), he is letting this subaltern suburb speak. Allied to this, Griffiths suggests that with this voice we also gain access to a worldview and knowledge of a way of life that is rooted in this place, as Robson confirms:

> "I kind of want to show [in my plotlines] that [life here] it's messy. In some ways its anti-social or it's maybe not at as polite as it should be. There's a great sense of humour here, there's an appreciation of life, there's a certain amount of set values which come about from living in a relatively poor area but it doesn't mean that people are stupid, but sometimes the quality of your Indian takeaway is more important than what's happening on the news, you know?" (PC)

Figure 4.3 The insider's view of Edge Territory. (Source: Author)

Robson therefore gives us the insider view of edge territory. He documents the present rather than memorialising the past, not giving the definitive account of the "the heavy streets. The hard streets. The hooligan streets" (Finch, 2003:21), but exploring "how we are in these gritty places and the ways in which we change" (ibid.). In Robson's words, mine "is an inside view isn't it? Portraying [life here] for what it is; not talking down about it just saying this is how it is for a lot of people, and there's no shame in that" (PC). In conversation I say that his writings give the reader a view of the world not unlike a documentary, with the reader/viewer seeing his world as if through a camera mounted on the character's shoulders. This ethnographic eye/I (after Ellis, 2004) resonates with geographical methods which seek to, "understand parts of the world as they are experienced and understood in the everyday lives of people who actually 'live them out'" (Cook and Crang, 1995:4). Robson's writing, to some extent at least, allows newcomers to edge territory to "understand th[is] world and ways of life of actual people from the 'inside', in the contexts of everyday, lived experiences" (Cook, in Flowerdew & Martin, 1997:127). That is not to say, of course, that these are works of pure autobiography. With their roots in reality, they are stylised and fictionalised accounts of life; as Robson puts it:

> "you know the true version [as you lived it] and then you kind of retune it [as a writer] ...yes a few things are tweeked, it's not 'real life' word for word . ...I talk about the drugs world and just the basic street violence, domestic violence, all that sort of stuff, lots of swearing. I am amazed by the reaction of some people in so much as they only see me as the character in the books rather than remembering that I am also the poet and so therefore have an element of intelligence, an element of sensitivity and all that sort of stuff... not that many people manage to put the both together [when they consider my writing] which I would think is the basic way to look at it really." (PC)

An excerpt from the short story *How Timex Tony Earned His Name* outlines the extent to which some Robson's narratives highly stylise life in edge territory:

> "Timex Tony got the nickname from his woman. And that was just about the last thing she gave him.
>
> After yet another night flirting and dancing and alley-shagging with some lonely, divorced, four-eyed temp called Tina or Marina or even bleedin Tia Maria, Tony arrived home late from the Tred, wasted, skint and spent.

> ...She raged and spat in his face, What time ju call this? and, before he could answer, she slammed [a] scalding alarm clock in to his face.
>
> Tony fell like a sack of spuds. His missus, towering over him, knelt on his chest and forcibly introduced Tony to the correct time, down to the minute, again and again. Fair do's, she's a hefty maid, and Tony couldn't escape. Ignoring his screams, she shoved the clock against his searing flesh, and ensuring the tea towel was between her and it, sat her bare cheeks on Tony and the clock getting face to face.
>
> Still, today, Tony bears the name 'TIMEX' branded on his forehead, to remind him, whenever he looks in the mirror, of the ravages of time. And the scars from the clock-hands warn every woman he meets, exactly what time he should get home to his missus. And these days, he's never late." (Robson, 2003)

The dialect and daily life of Robson's area of the city is unique. The sense of place in this particular territory is, to an extent, an illustration of the hyperlocal. The hyperlocal refers to information oriented around a community with its primary focus directed toward the concerns of its residents. The hyperlocal thus alludes to the communities at the scale below the local, to the intimate and everyday goings on in a small number of streets across a neighbourhood. A sense of community is generated through this street life, and a cohesive sense of identity that to some extent is insulated from those outside. As Robson states:

> "I can walk round Canton [an example of a neighbourhood in Cardiff] and I kind of know the place, I know my way geographically but it's different form of knowledge as we've never been a 'Canton family'. We've always been over here. My aunty and cousin still live on Stacey Road, some of them have stretched as far as Cathays. Yes, I have family up in Ely but they're kind of distant family [both geographically and in terms of relations], even though it's only a few miles away that's over the 'other side', you know? There is a geographical split with the river [in Cardiff] and anything that's 'over there' is as good as being emigrated. I mean ...the absurdity of that boundary is extremely real still for a lot of people. Not so much in the middle classes, but in the working classes there is a line. Whether that's cultural or just a sense of territory. Obviously there's a sense of scale in Wales, we're a small place. You can live a mile away and that makes you a sworn enemy you know because you are one of 'them'. We are very tribal and territorial and that tribal nature is still here." (PC)

As Robson implies, every hyperlocal has its own (hyper)locale and sense of place. The hyperlocale can be understood as the built, natural, and social environment generated by cultural relations on the streets – it is the composite of all the actions and practices that come together in one place. The hyperlocale therefore provides the setting for everyday routine and social contact – in the words of O'Loughlin & Anselin (1992:16), the hyperlocale "is the setting of interaction and the contextuality of social life". As such, the hyperlocale comes close to identifying the character and feel of a place. Allied to this, the hyperlocal sense of place refers to the emotional, experiential and affective dimensions that tie humans in to particular environments. Through genealogical or generational linkage, childhood memories, or key cultural events, our personal and emotional lives become intertwined with the hyperlocal (see below for examples).

Robson's hyperlocal is to an extent bounded in nature. As we have seen in Chapter 1, many geographers frame place as being bordered in nature, with clear identifiable boundaries which serve to demarcate 'inside' and 'outside' territories, and also serve to identify 'them' on one side, and 'us' on the other. Although this bordered nature of Edge Territory is clear in terms of the physical boundaries of the River Taff (between East and West Cardiff), or even to a lesser extent by the imposed boundary of Newport Road (between Splott and Adamsdown on one side, and Plasnewydd and Pen y lan on the other), Edge Territory remains a difficult place to cartographically locate. As Robson outlines:

> "Well I suppose kind of one of the things I like about this area is you could argue that its right at the tail end of Roath; you could argue that it's the tail end of Splott; you could argue that it's the tail end of Adamsdown, and I like that. Historically this is Roath but then historically a lot of Adamsdown was also Roath so where the boundaries are change over time and I quite like that. There is an ambiguity there so you can make the place your own, absolutely." (PC)

Figure 4.4 The mysterious location of Edge Territory.
(Twitter feed and Author Photo)

“roath splott adamsdown. not a triangle renowned for its miracles not a triangle at all; a linear grasp a rasp of frictional existence & tender hammered attitude, i am from none of them, yet all. my family from adamsdown & tremorfa; adamsdown when it was part of roath. splott what i get accused of like it’s something to be ashamed of by those

> from the city's west or north, this side of the bridge so not splott at all, but within the call of the splotlands' last orders a pint of dark don't mind if i thanks very much ta iechyd da 'ware teg down the hatch & up yours butty bach you-arr en shovel it down ya neck then pal." (Robson, 2003:95)

The Roath-Splott-Adamstown triangle appears to defy convention. Noone is certain where one area begins and the other ends, and in this ambiguity, as Robson acknowledges, there is a freedom to be yourself. In many senses, therefore, this area of the city is a 'third space'. Thirdspace, as postulated by Soja (1996, 1999), is a deliberately flexible and tentative concept that seeks to open up alternative ways to comprehend "the meanings and significance of space" (1996:1). As I have argued elsewhere (Anderson, 2002), it is a response to what scholars such as Bruno Latour (1993) have identified as the 'modern constitution' which creates 'Big Dichotomies' (Soja, 1999:268) or 'either-or' categorisations (such as subject-object, nature-culture, man-woman, black-white) which limit our understanding of the world. In Soja's words, third space resists:

> "any attempt to confine thought and political action to only two alternatives, by interjecting an-Other set of choices. . . . The original binary choice is not dismissed entirely but is subjected to a creative process of restructuring that draws selectively and strategically from the two opposing categories to open up new alternatives." (1996: 5)

Life in the edge territory of the Roath-Adamstown-Splott triangle can be understood as a third space as it cannot be located simply in one of the bounded neighbourhoods, it exists in the ambiguous space between them, drawing on each of their locales and sense of place to define a new alternative. In this thirdspace Robson sought to look for his own identity, and perhaps due to the ambiguity and otherness of this space, was able to find his own alternative voice and sense of belonging. Like many others in the city, Robson is not *from* Cardiff, but he is now *of* Cardiff. With routes leading him from Plymouth, Monmouth, and Cwmbran, amongst other locations, his writing is now rooted to the city. His voice is now a Cardiff voice, or perhaps more appropriately put, a *Kairdiff* or *Kahrdiff* voice, which attempts to articulate his story of trying to find where he came from, and where he now belongs, in both a social and geographical sense.

> "i could so easily claim elsewhere, so easily claim cwmbran, monmouth, plymouth, herefordshire – each accepts me as one of their

> own. i could so easily go elsewhere could so easily claim british the empire state ich dien ich bin ein european i am of the continent introduce myself as a citizen of the world community the globe the planet could so easily be permanently mentally abroad & out of here cut loose from static.
>
> sometimes i do, i describe myself as such, sometimes as welsh, more often just Cardiff, kahdiff. more often than that: roath or splott or adamsdown – a delicate answer depending upon the city-grasp held by the listener the questioner the other side how precise do you want me to be exactly ay? come on now gimme a chance, place of birth: st david's hospital, on my birth certificate Cardiff gets mentioned eight times; wales not once." (Robson, 1993:95)

Here Robson identifies the initial complexity of answering the apparently straightforward question: "you kairdiff? where ju [live]?" In this excerpt he acknowledges the diversity of scales to which we can form a sense of belonging, be it the global, the continental, or the national. Humans can therefore enjoy a nested sense of belonging at multiple scales, but at the same time we can also answer the question more specifically – we can be from a town or city, a neighbourhood, or even from the street where we were born. Robson has become attached to the thirdspace of his particular hyperlocal neighbourhood. Robson writes his journey to find his place in this thirdspace not through the lens of a collective history (like Finch), or an imaginative characterisation (like Williams), but rather gives a (semi-) autobiographical account of locating home in edge territory. Like American writer bell hooks (another adherent to the radical alterity of the lower case), Robson wanted to: "find my place in this world, to have a sense of homecoming, a sense of being wedded to a place" (2008:1). As Robson writes:

> "perhaps we had it once: our family history, perhaps we couldn't afford its upkeep, pawned it, swapped it, drowned it down the toilet like me uncle john's parrot before he went back to practising the trumpet, dropped it getting off the bus like a baby's spouted cup full of warm tea, got drunk & left it in the pub, stashed it up the chimney in the newly weds' flat, perhaps it was sawn into strips & lengths & planks at a sanquahar street timber yard or buried under the parky's house in a llanrumney public park, perhaps it's been warped played down exposed mutated mutilated exaggerated to such an extent that we have forgotten what it was we were talking about, perhaps we ate it in a fit of munchies, perhaps my invisible friend took it with him, perhaps the old pissed on it cried on it died on it rested upon it '*just half a cup lil'* spilt it chucked it over the side on a £10 one-way ticket

> to australia commissioned it enlisted it signed it up got it under contract press-ganged it into service launched it floated it sunk it depth-charged bombed torpedoed it perhaps we tugged it out of the dock perhaps it is swimming somewhere or lies at the bottom of the Severn chained to a rock perhaps we drowned it at birth tied it up in a sack aborted it before it had chance perhaps it is still to be fertilised still ovary & sperm an excuse to get down sweaty & turned on spent & done, perhaps clone or test-tube borne, perhaps it lies in the tip of a condom floating upon roath park lake waiting to choke a pintail a pochard a shoveler a wigeon a tufted an escaped ruddy shelduck or down the canal with the bridge-livers the taff with the jumpers the waders the face down drunk perhaps we couldn't bear it at all perhaps the weight of it broke our back perhaps it is left in a doorway somewheres to be brought up by fosterers front page of the echo & renamed by staff perhaps it puffs at a mouth organ & sits by a hat perhaps it is shafted on scrumpy jack perhaps we are running running running from it fast as we can perhaps it is hiding waiting to jump us when we're fool enough to relax & consider the future ours, clean pure of its influence in blue skies calm seas & sunrise yes, perhaps it is that.

"me: i hunt the city for it." (Robson, 2003:97/98)

Robson's search for origins and belonging are not simply complicated by the scales to which he can identify. We have seen above that he can claim citizenship at a variety of scales, but if we take the question "you kairdiff? where ju live?" seriously, we can ask how far we should go back in time in order to qualify our answer, as well as how specific we should be in terms of spatial scale. When we define our identity is the issue of where we currently live more relevant than where we were brought up? Is a sense of emotional belonging more important than our birth certificate or passport? Perhaps we could even consider where we were conceived, or where our parents were born as relevant complications to our answer. Belonging, it seems, just like the border of the Robson's hyperlocal triangle, is not easily drawn. As Robson states: "i am a Cardiff boy but this does not encapsulate does not define completely myself nor my experience we are both more, signed myself up for a lifetime of explanatory notes…" (2003:96). In complicating identity in this way Robson has created a form of 'third space' for his own sense of belonging. By siting himself in a space which "eludes the politics of polarity" (Bhabha, 1994:1) and finding a complex, ambiguous space in which to call home it has become possible for Robson to "elaborate [alternative] strategies of selfhood, [and] initiate new signs of identity' (after Bhabha, 1994:1).

Robson's origins are therefore complicated, yet what is clear from the above excerpt it that all the key events in his life were co-ordinated somewhere. In other words they all 'took place' and these geographical coordinates claim a fragment of his history and identity. We have seen above that Robson's family is not a 'Canton family', and through hunting the city for his origins, he came to realise that his extended family has both roots and branches throughout this hyperlocal thirdspace:

> "Well as an example I used to live in Ruby Street off Clifton Street. Now there's a lovely series of roads that go off Clifton Street, it goes Sapphire, Emerald, Ruby, Topaz and Pearl etc. My grandfather lived in Ruby Street, my father lived in Emerald Street, I then later on lived in Sapphire Street and also Ruby Street, ...I can trace my family from three generations [on these streets] which is a millennia in Cardiff terms." (PC)

The Robson family jewels could therefore be deemed to be geographical places in this hyperlocal. The gemstones in the streetnames of this neighbourhood were "the only geographical historical cultural & personal constant" (2003:101), and have become touchstones for his memories and key events in his life:

> "[i] find myself blasted to the past walking down streets turning corners looking up or down, kneeling to tie a shoelace being faced with a perspective not experienced since childhood getting it all come flooding back, mental & emotional explosions relived in the city, ourselves & our mutual pasts incessantly & irresistibly layering us folding us under a blubbering belly of insight & experience as we stomp around headphones on or plastered drunk or sober & in a rush suffocating & breathing life reminding us of fights & piss ups & snogs & lust, humilities & hangovers, dancing swinging swearing reading writing on lamp-posts swearing & loving & sulking & kicking off *what the fuck did you mean by that?!* crying in the streets on the way back to someone's house if not my own trying to apologise trying to make up. child & adult, not much different after all this time, knowing everything, reading the footprints, the smells in the wind, the hang of the stars, finding myself & getting lost." (Robson, 2003:102)

Through this process Robson realised the integral role of place in the construction of his identity; in the daily activities of 'incoming' and 'outgoing' to and from locations (see Casey, 2001, and Chapter 3) the city became central to his geographical awakening (see Box 4.1). Through claiming his streets, he realised how:

> "Human experience is formed and gleaned, life-sharing managed, its meaning conceived, absorbed and negotiated around places. And it is in places and of places that human urges and desires are gestated and incubated, live in hope of fulfilment, risk frustration and are indeed, more often than not frustrated." (Bauman, 2003:102)

Box 4.1 Bittersweet belonging: Life absorbed, negotiated and frustrated in places

In this hyperlocal edge territory Robson hunted for where he belonged. However, following Bauman (2003), he realised that rootedness in a place not only provides connection and belonging, it can also provide a sense of claustrophobia and constriction. In finding home one also loses a sense of absolute freedom, as the positives related to being bound come with a price.

> "i chopped down every tree i saw in search for it, kicked over wheely bins, swore down alleys, promised i'd get it, uprooted floorboards, spat down cellars, hid in attics, broke windowpanes & scattered the shards & fragments poked my nose into graffiti'd lockups sprayed my stench along the crumbling plaster walls so it could follow my scent know i was on its tail its case on the look-out so it could catch me up & hunt me down ...i telexed telephoned telegrammed emailed wrote letters postcards took photographs collected party snaps & weddings framed by confetti & the steps of a church i faxed fucked & became well fed always staying for tea well while i'm here i may as well i built up the evidence my case file was bulging i was cutting the forest pressing the paper & creating the backlog & eventually, when the city was bare & all was there for me to see & feel & grasp between my sweaty grabbing mitts, the very worst happened: i found it. or it found me. i grinned at it hid from it towered overshadowed me & now it's hounding me chasing me & i have nowhere left to hide for i chopped down everything in search of it & now on the concrete the tarmac it has clamped me & i am immobile i am whip-tied lashed lasso & lariat & it drags me back in its roy rogers t-shirt it drags me down down o clever me, o driven me, o me in search of myself & my city, now it has

> found me what do i do with it what does it do with me? for my debt for the debt of my family i repay with interest upon interest compounded in bankruptcy for the rest of my lifetime, my search, my family, my history, my city, i had to go find it didn't i; wouldn't leave it be. dozy fuckin idiot, me me me." (Robson, 2003:98/9)

In experiencing the bittersweet nature of belonging, Robson identified how the place of Cardiff had come to function as a surrogate family in his life, with all the tensions involved in that role:

> "Cardiff was is a surrogate dad. father & son aren't meant to always stand shoulder to shoulder see eye to eye without some kind of poker stare friction some kinda punchy rub some kind of man to man don't push your luck what you staring at a desire to prove something to each other the worm that turned you might have your wind up but i'm still your old man don't you go forgetting that lad." (Robson, 2003:103)

Extending familial connections from bloodlines to plotlines resonates strongly with the notion of 'transpersonal ecology' (Fox, 1995). According to Fox, the process of transpersonal identification involves not simply viewing our identity as defined by our selves or "our family, ...friends and distance relations" (Fox, in Gottlieb, 1991:414), but also by our relations to places. Fox describes this geneology by comparing it to the roots and branches of a tree:

> '[this process] leads us to identify ourselves more and more with the entire tree rather than just with our leaf (our personal, biographical self), the leaves on our twig (our family), the leaves we are in close proximity to on other twigs (our friends), the leaves on our minor sub-branch (our community) ... and so on. At the limit, [it leads us] to impartial identification with all particulars (all leaves on the tree).' (in Gottlieb, 1991:417)

It rooting himself, Robson extended his family tree from his blood relations to the plotlines that criss-cross the edge territory of Cardiff.

"A space of radical alterity. enter that space" (hooks, 1991:152)

As stated above, in Robson's search for belonging he gives an insider view, and also an insider voice, to his hyperlocal thirdspace. Despite the tensions inherent in this new found belonging, Robson displays loyalty to these roots and the influences it has had on the way he speaks.

> "my accent is a shambles – i admit it readily; a just reflection of my loose view, it is the only map to my existence but there are too many contours leaving listeners confused, i will not deny my range my diversions; one minute Cardiff st german's, the next plymouth st jude's. i've confessed to both in my own voice so they deserve to be heard." (Robson, 2003:96)

In this defence of the polyvocal nature of his accent, Robson exhibits a political element to his writing.[1] This dimension of Robson's work is articulated in part simply through his deep seated desire for freedom of speech. Originating from a childhood where a stammer affected his spoken word, he is now assertive in everyone's right to have their voice heard. As he states:

> "I kind of went from not saying a word to being a mouthy bastard really, [writing gave me] a new found freedom and it was like okay well this is my voice and if you don't like it fuck off ...don't try and stop me having a say because I'll kill you for it." (PC)

This defence of the freedom of speech, and in particular local modes of articulation, are also manifest in his defence of his own version of the Cardiff dialect. Pointedly, Robson does not share the view that the Welsh language ought to be the lingua franca of the country (see also Davies, Chapter 10), but rather retains his valorisation of the (hyper)local. The conventional position is that the Welsh language (or

---

1 According to Peter Finch, Robson is "a post-Poll Tax scribbler" (Finch, 2003:21) , and this more overtly political aspect of his writing was noted in the preface to 'Edge Territory' (1995) where Robson rallied resistance to the latest legislation that sought to enclose peaceful assembly, freedom of movement, and rave parties (see McKay, 1998): the Criminal Justice Act (1994):

"insist / on your personal freedom / mauled again 03/11/94 / keep fightin / the criminal justice act"

Cymraeg) has been marginalised and largely silenced by the English language (see Chapter 9, and Chapter 10). However, Robson argues that Cymraeg is bound up with its own forms of colonisation: it too seeks to enclose and dominate sub-national dialects and languages in order to create a unified Welsh voice.

> "The issue I have fundamentally on Cymraeg even though I love it, great language, I refuse to be forced to speak it, or refused to be forced to feel guilty for not speaking it. I speak English in a way that only occurs in Cardiff and Cardiff is in Wales and therefore I speak Welsh. I may not speak the Cymraeg language but I do speak Welsh. Cymraeg activists would have you believe Wales only had one language and it never did. It had at least three [before you even consider dialect]." (PC)

As he writes in *A Stain on the Map*:

> "i have reversed my father's attitude, as with so many of my parents' age, they fought to lose their Cardiff accents to become acceptable in vocal range angle & rasp, we are regaining it on their behalf, i doubt they will appreciate this so i will explain to those dead & alive: Cardiff today does not need to be ashamed of how it speaks or its many accents, the city will begin to accept itself for what it is before it can expect others to value it. we have a magnificent range of voices & accents, the beauty of how we think & speak, communication, after all, should be about what is said & not how. to judge a speaker on their accent, to base their value their class on how they sound, to balance whether we listen or not, to create abstract hierarchies based on vocal intonations & musical patterns, is to create inefficient ineffective irrelevant communication & true dysfluency – but on the listener's part, is to insult, my parents did not realise this, nor did those so many of their age who did & continue to do the same, they simply wished to play on a level playing field, so do i, but home not away, there is no need to unnecessarily limit the kahdiff experience; do not deny us our claims to diversity, we got range – it's the modus operandi the success the style the soul the truth, ruth, the kahdiff accent a myth perpetuated by those who would distance us from the couth – if you reckon they talks the same in ely as they duzz in tremorfa you're revealing your ignorance your ear ain't in tune, we got it & we got every right to keep it. it ain't english, it ain't cymraeg, but iss kahdiff welsh, true." (Robson, 2003:96-7)

In his willingness to fight for his voice, Robson questions not only the simplicity of roots and the tendancy to affiliate to a singular scale (often the national), but also champions the view that the quality of your local takeaway is often more important to every life than what is on the news. In challenging the 'imagined' singular voice of Wales,

Robson celebrates the diversity of his own roots, and those of his hyperlocal, seeking to articulate how this diversity should not be silenced; in his words:

> "the world we live in is changing, [and to those who seek to tie us to any one version of Wales I say] stop fucking making us smaller by being so stupid and fucking restrictive on what is Welsh and what makes us Welsh. You fucking fly to space it's the Welsh thing to do, you know, you go and fucking snorkel bogs, it's the Welsh thing to do, whatever you do you are Welsh, you fucking decide you know?" (PC)

From this faith in the diversity and radical alterity of his voice, hyperlocal, and identity, Robson also offers a defence of English as a language. He does so by framing it as an important resource that can be pragmatically used to articulate ideas across the globe, but also as a means of communication that can be indigenised and colonised to suit particular local places, as he states:

> "...There's lots of people in Wales who hate the English language simply because its called English but it enables us to communicate throughout the world so it's actually a fantastic tool. Why should we hate a fantastic tool that enables us to communicate simply because it [was] English, or it's called English? We should be proud of [our version of English.]" (PC)

Robson implies that in language, as in geography and identity, a thirdspace position can be taken that opens up the possibilities for creation and articulation. By moving beyong binaries of either speaking Welsh (Cymraeg) or English (in terms perhaps of Received Pronounication), we can see the 'interjection of an-Other' set of voices (after Soja, 1996). This may be a Roath-Adamstown-Splott dialect, a 'kairdiff' voice, or even a hybridised version of Welsh and English (or 'Wenglish', see Elmes, 2005). By exploring these linguistic 'third spaces' it is possible to find a voice where neither Welsh or English "holds sway" (Routledge, 1996: 400) and we consequently "emerge as the others of our selves" (Bhabha, 1994: 39).

Cardiff Cut: *"...locking on... travel pak... broadway orbit... edge territory... zone no control.... traffic snarl"* (Robson, 1996: no page)

So far we have argued that Robson is a poet of 'edge territory', an occupant and chronciler of a geographical thirdspace in which location, language, history and identity do not fall easily into modern categories. Instead edge territory as third space opens up new possibilities for both geography and fiction as it enables new combinations, spaces and possibilities to be imagined and inhabited. Robson's Cardiff is thus cut from the fabric of city but redesigned in a new way beyond the control of traditional lines of custom or culture. It conjures up a version of the city from the perspective of a 'third-spacer', reflecting the experiences of a Generation X whose lives are defined in different ways from those inhabiting the regenerated spaces of Cardiff Bay, or the suburbs of the city (see Chapter 5). This generation lives their lives on the bejewelled streets of edge territory.

Figure 4.5 Cutting the plotlines. (Source: Author)

"To the south, the streets of Splott and Tremorfa are hampered by a dense corrugation of speed humps that slow even the Kawasakis that leap across them. Road deaths in poorer districts were long thought to be the fault of addled youth spinning sparks out of the road surface in their side-skirted, sewer-piped Peugeots and speaker-stuffed Novas. Research has shown that they are more a product of the amount of time people here actually spend on the streets and the number they need to cross in order to get where they are going. Still, nothing quite like seeing a gleaming boy racer fingering his earring as he boomboxes along at a five mph crawl." (Finch, 2004:15)

Figure 4.6 Street life in edge territory. (Source: Author)

Finch's description sums up the importance of the street to edge territory. Mathews et al (2000) use the term 'the street' as way of encompassing the places in which some cultures, particularly youth cultures, claim to articulate and express their identities.

> "We use the term 'the street' as a metaphor for all public outdoor places where [some cultural groups] are found, such as roads, cul-de-sacs, alleyways, walkways, shopping areas, car parks, vacant plots and derelict sites. ...To [some groups] the street constitutes an important cultural setting, a lived space where they can affirm their own identity and celebrate their feelings of belonging. In essence, these places are 'won out' from the fabric of [mainstream] society, but are always in constant threat of being reclaimed." (Mathew et al., 2000:281)

I would argue here that the streets of edge territory are an important space for the cultures Robson depicts in his fiction: the streets are 'won out from the fabric' of the mainstream, and through cruising in 'sewer-piped Peugeots' and watching 'speaker-stuffed Novas' the streets become a stage on which lives are lived.

However, living and being mobile on the streets of edge territory is not only practised through pimped-up rides, but also on foot. In *Cardiff Cut* in particular, Robson (2001) creates a plotline from the city centre into edge territory from the perspective of a walker through the city, engaging with the city from a perambulatory perspective. This choice of engagement I would argue is significant. Not only does it emphasise the importance of the streets to the life of edge territory – the walker does not travel through an abstract space (in a bus or taxi for example) and thus is not insulated from the smells, people, and atmospheres of their environment. Rather they directly engage with the gradients, paths, obstacles and materials that come together to create a place. Walking is thus an act that creates a different experience of the city. As Morris discusses:

> "space and place are not merely inert or neutral features of the built environment; instead, they must be activated by the 'rhetorical' practices of users and passers-by. [As de Certeau] comments: "'if it is true that a spatial order organizes an ensemble of possibilities (e.g. by a place in which one can move) and interdictions (e.g. by a wall that prevents one from going further), then the walker actualizes some of these possibilities. ...he also moves them about and invents others, since the crossing, drifting away, or improvisation of walking privilege, transform or abandon spatial elements' (1984, p. 98)." (Morris, 2004: 677)

Walking is thus an act which can challenge dominant uses of the city and, by extension, pre-existing understandings of these spaces. Due to these possibilities, for Bairner (2011:382): "walking teaches us more than any other activity about the places where we live and the places that we visit". In answering Virilio's question "can asphalt be a political territory?' (2006:30), Bairner answers "an unequivocal yes" (2011:376). In a sense then, walking can be seen as a transgressive or political means of mobility, as Macauley tells us, walking can be seen as a "peripatetic politics" (2000:3). The importance of this peripatetic politics for carving out a sense of belonging cannot be underestimated, as bell hooks notes:

> "I need to live where I can walk. I need to be able to walk to work, to the store, to a place where I can sit and drink tea and fellowship. Walking, I will establish my presence, as one who is claiming the earth, creating a sense of belonging, a culture of place." (2008:2)

In the same way that Andrew Craig Williams created his sense of place in the city through everyday incoming and outgoing on foot, Robson depicts a peripatetic understanding of the city through his plotline in *Cardiff Cut*. *Cardiff Cut* is a plotline from St Mary's Street, through Mill Lane, Caroline Street, back to a house party in Splott, then along City Road home. In this story Robson offers us a thirdspacer's perspective of edge territory, of lives interested in drugs, alcohol, sex and violence, which moves through streets to change our understanding of how city life is lived.

> "'cold and tired pop in relax have a nice cool drink'

(windowpaint, spielothek amusement arcade, prince of wales theatre, st. mary street).

straight to the front of queue girls tryna get ina philly, lines of boys under lion canopy pissing their money over each others' shoes not a long sleeve between em not a goosebump let loose." (Robson, 2001:10)

Figure 4.7 St Mary's Street 'Philly and Caroline Street Alley. (Source: Author)

"need food, consider a Chinese but remember lastyear's poet in chowmein free-for-all incident/fraudulent use of/getting chased up the street, decide against it. settles for dorothy's fish bar in Caroline street, cross & pass the shoepurmarket/taurus steak house/charlestons brasserie (an arc of chips surrounds a hit&run victim where Caroline meets st. mary: not so much *cashed them in* as flung em each end of the zebra crossing), trudge the masses pisst up & in out up against wall of brains brewery: a mark of respect for the nite's liquid intake, otherwise known as recycling." (Robson, 2001:10)

...

"city road:

flyposters announcing

'late nite flava' *(i see* phantasia'*s shut again)*

a stagger of moonbeaters pass the 9-til-5's not fusst about finding ways home/too busy filling face wi burgers, kebabs, & *kinell!* a bag fulla donuts *(stay away from the hole).*

city road:

helicopters tracking video pirates & raving airwave buccaneers careening careering stolen cars with stereo at maximum/ handbrake turns into

bedford street:

campervans wired up to the neons (free electric thro the nite, a dodgy gas canister leaks on the quiet, come morning: a fifty-a- day-man stops alongside, sparks up & takes half the street sky hi. for a whole day it'll rain ikea & habitat: bedford refurnished, sodium bright)

some kid leans & shouts outa upstairs window

"mayday! mayday!" mother screams from neighbour's bed

"get ta sleep ya little bastard or ill takeya mobile away"

get back to the flat, the girls walk with us to frontdoor, one sez

"we'll come in fora cupa but we're goin home to bed" "yeh, right, yeh..."

we go up a flight, make sarnies, a cupa & smoke; a table fulla teastains/lotsa little blimburns/empty bottles/full ashtrays/ discarded filters & roach, i take off me hat&coat, settle in to a comfy seat & remind meself of this cheeky little mexi. begin the first of many searchpatrols for the telly remote.

the girls stay a while, one writes down a number which she gives to me mate, have forgotten their names by the time they vacate (i ain't renowned for remembering, but they was good companions while we paralleled trips)." (Robson, 2001:51/3)

...

"newport road:

takes a piss in a shagbattered doorway, aware of it cold but no feel it juss steam, the old lloyd's bank sprayed

'jacks' & 'scfc'

> tho long painted out can still read it beneath (cheekybastards, if you looks close enuf it becomes unremitting/at this time of morning: a smack in the teeth)
>
> & i'm stopping&starting & stopping&starting like the stations of the fuckin cross *(i thought you wuz ina rush?/get homeya daft cunt...)*
>
> ...
>
> end of roll." (Robson, 2001:68/90)
>
> ...

Conclusion: "my only compass broke & stumped & set to the east of queen street cobbing the hot dog of newport road" (Robson, 2003:97)

In this chapter we have seen how Robson uncovers the importance of geography in forming his identity. Through hunting for his sense of self in the city, the external landscape of edge territory has got under his skin and become internalised into who he is. For Robson, his identity is formed in the gaps, ambiguities and tensions of edge territory. In the third space of the Roath-Splott-Adamsdown triangle, Robson seeks to uncover his roots and wrestle with the responsibilities of being tied to a place. In exploring this edge territory, Robson offers a new take on belonging and the city. In the words of Soja, Robson discovers how his plotlines are, "filled not only with authoritarian perils but also with possibilities for community, resistance and emancipatory change" (Soja, 1996:87). He demonstrates how it is possible to resist inherited notions of family or nationhood and tie ourselves into new spaces which, "give rise to something different, something new and unrecognisable, a new area of negotiation of meaning and representation" (Bhabha 1990: 211).

In exploring this new space Robson revels in the complexity and contradictoriness of his constitutive coingredience with place, as he puts it: "i refuse to accept my social limitations i refuse to be tied to being only one person i am a myriad a myriad my raid a myriad i am" (2001:90). The potential for individuals to have a complex relationship with place will be explored further in Chapter 5 (as well as Chapter 12). For now, however, our plotlines leave edge territory and move north to the more sedentary suburbs of Roath and Roath

Lake, and in these more stable locations we see a new set of characters experiencing a different set of person-place relations in the city of Cardiff. As we do so, we leave Robson continuing to undercover his ongoing relationship with place and its perpetual influence on his character and life:

> "this city breaks my heart & holds my hand, rubs my nose in my own filth my own past my own extremes & limitations my own limp dysfluency my inability to speak to articulate communicate my own bad sound, it haunts me bullies me slaps me about, it pulls me around gives me Chinese burns when i ain't playing no more & wanna go & sulk, it puts me down humiliates me in public but then, when no one's looking, drags me back up, strokes my hair, gives me a clip round the ear, tells me i deserved that now stop being such a little tit & get on with life, be a man.
>
> ...Cardiff as rumbling creative stove & store? too fuckin right, take me back." (Robson, 2003:101/104)

Chapter 5
Tessa Hadley's Roath: the meeting place of suburban dreams

> "Parks are places where you can sort your problems, [they are] the epicentre of suburban dreams." (Finch, 2002:35)

I'm sitting in 'Terra Nova', the park café which looks out across Roath Lake, Cardiff. The café is named after Robert Scott's ship which sailed from Cardiff on his fateful expedition to Antarctica in 1910. These global routes play across my mind as I sit with a large black Americano, looking out on an August afternoon whose skies are masked by a "grey-green cocoon of rain falling" (Hadley, 2005:91). As I gaze through the rain a solitary pedalo crosses the lake and scatters birdlife – gulls, swans, ducks, and moorhen – who were lazily scouting the surface for food. An old man, peak capped, carrying a plastic bag full of bread, emerges on the eastside of the water. The prospect is enough to alight the colony of doves from their idle roosting on the café roof.

Figure 5.1 Terra Nova and Roath Park Lake. (Source: Author)

It is here, or an entangled version of here as I will later explain, that Tessa Hadley set a key scene from *The London Train*. This novel, among a number of short stories by the same author, uses the suburb of Roath as a key stage for its plot. *The London Train* is a novel formed by two characters whose lives entangle for a short time following a shared train journey. Unlike the characters in Williams (Chapter 3) or Robson (Chapter 4), Paul and Cora are in their early middle age. Paul is a writer and originally from Birmingham. He left his childhood roots when he was young, and now lives in the Monnow Valley with his second wife and two young children. Paul also has teenage daughter from his first marriage, but she has moved away to London. Cora was born and bred in Cardiff, and works from a city library helping young people with reading problems. She is separated from Robert, a civil servant, and she lives in a house in Roath. On the surface at least, Paul and Cora appear settled and happy (only Cora's close friends appear worried about her). However, both have recently lost their respective parents, and this episode has reminded them of the passage of time and emphasised they can no longer consider themselves to be children, or young for that matter, any more. Both stories chart their respective character's progress in attempting to find where they belong in their 'middle age'. It is the time in their lives where they assume they ought to have arrived where they want to be, that they should feel as if they belong. However for both characters life has not turned out in the way they may have planned. As Hadley writes, both characters seem to have "fallen out of a very intricate piece of machinery which had been [their] life' (2005:93), and echoing bell hooks, they both feel 'out of place' in their superficially successful lives:

> "Talking about place, where we belong, is a constant subject for many of us. ...I am stunned by how many citizens... feel lost, feel bereft of a sense of direction, feel as though they cannot see where our journeys lead, that they cannot know where they are going. Many folks feel no sense of place." (2008:1)

Hadley's characters Cora and Paul meet on the London Train, and to echo this literary device it is on a train I meet Tessa Hadley to hear her discuss her novel. It is notable that her characters, unlike those in Williams (Chapter 3) and Robson (Chapter 4), do not meet whilst walking their city, but rather meet on a train. Their train and ours, journeying in and out of Cardiff, evokes the movement, displacement, anonymity, and dislocation of identity that Hadley's plotlines engage with. The place of *The London Train* reflects their lives: it moves through a landscape but is disconnected from it, it travels at a dischordant speed, with its passengers simply observers as the world passes them by. Rather than the engaged, politicised walkers that Robson narrates, the train encourages "a kind of self-absorbed 'sleep walking'" (Macauley, 2000:3) through the landscape, and perhaps by extension through life. The train is inhabited by strangers, detached from one another by various 'involvement shields' – by newspapers, sunglasses, and mobile phones (see Davidson, 2003) – where individuals have to suffer other's noise, but also have the opportunity to encounter new people. In some senses, therefore, the train is an anonymous in-between space: it is the ideal space for rootless people to meet.

Figure 5.2 Tessa Hadley and The London Train. (Source: Author)

In some senses, we can understand the space of the train as a liminal space. Liminality, according to van Gennep (1960), traditionally refers to a position on the threshold or margin at which activities and conditions are most unstable and uncertain. Liminal spaces "literally occupy… a threshold or borderland space, straddling, subverting, and disrupting boundary lines of separation and contact" (Jones, 2008:17). As Turner states, in practice liminal spaces and liminal beings are "neither here nor there; they are betwixt and between the positions assigned and arranged by... custom [and] convention" (Turner 1969: 95) and we might add geography. In Bhabha's terms, people in this space are "neither one nor the other, but *something else besides, in-between*" (Bhabha 1994:224, emphasis in original). Although often used to describe youth cultures – those individuals who are no longer children but not quite yet adults – I would argue that liminality can also be used to describe characters in their middle age: those who feel they are not yet 'old' (or even fully 'grown up'), but can no longer legitimately describe themselves as 'young people' anymore. As a consequence, these liminal people may have a peculiar affinity with the liminal space of the train. Train journeys occur betwix an origin and a destination and for the most part are located in-between stations; it is in this geographical and temporal in-between that both ennui and escape are experienced.

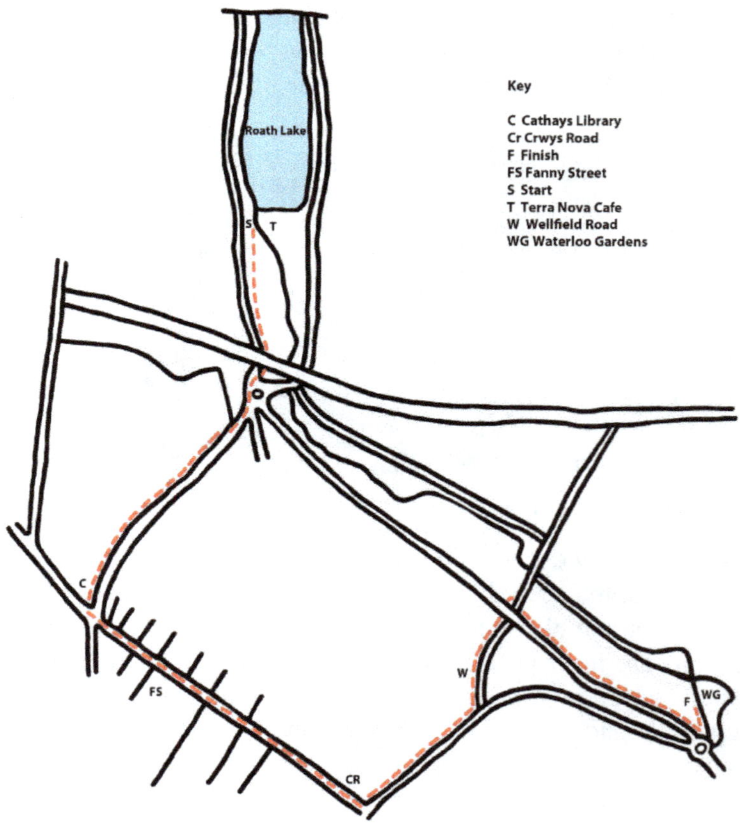

Map 5.1 In Hadley's Suburbia

In this chapter we examine Hadley's characters whose lives meet and part on the liminal space of the London Train. Out of place geographically and socially, both Cora and Paul seek a sense of belonging in their middle age, at once seeking escape from their past and present, as well as the chance to relive lost youth. We shall see how Hadley's work uses geography as a metaphor for inner life, and how suburban communities offer proximity without roots – that the dream of suburbia is not always as it seems. The chapter outlines the plotlines of Cora, then Paul, then the places where their relations are formed. We start at the Terra Nova café...

Cora's Plot

The Terra Nova Café is located at the opposite end of the district that descends into Robson's Edge Territory. This end is the suburban part of Roath, an area that is "given over to cultivated beds, bowling greens, [and] a rose garden" (Hadley, 2012:183). According to Hadley, it is a place where couples choose to spend an afternoon in the park with "baby-blankets and cushions, ...plastic cricket ball and tennis ball, a bottle of rosé and glasses... [as if they were] an idyll from [an] of old-fashioned children's book" (Hadley, 2012:183). Roath Lake is itself old-fashioned, it was constructed in the late nineteenth century on land owned by the Marquis of Bute, as Finch explains:

> "Bute knew that in order to build and sell high value property you need[ed] to provide amenity. In 1887 he offered 103 acres of upper Roath to the Corporation, got this matched by 18 acres from Lord Tredegar, and established the chain of public spaces that still run in a green line out from the city's heart. ...It took the Corporation ...seven years to dig, drain, pave and plant this Victorian splendour [and] ...over the years the Lake developed its own mythology." (Finch, 2002: 33/4)

Tessa Hadley's plotlines entangle themselves into this mythology. She integrates the 'Victorian splendour' of the past into her present, but like some of the grand houses on the lake, her characters have become fragmented, disjointed and disconnected; as she writes in 'The Trouble with Summer':

> "These are big [old] houses, you can [now] get four, five or six lets fitted into one of them. The place I'd moved into that summer had been the family home of an old lady who died; ... It was papered in a

> faded pink and green trellis, the lined silk brocade curtains were so brittle they were crumbling away under my hands, and I could see the lighter shapes on the wallpaper where the old lady's bedroom furniture used to stand." (Hadley, 2005: 92)

Figure 5.3 The houses on Waterloo Gardens, close to Roath Lake. (Source: Author)

Like the metaphor of the train, the Victorian houses and the suburban dreams they were built to home have decayed and crumbled, the inner spaces of Hadley's characters come to be reflected in the built form of this part of the city. For Hadley's characters the 'trouble with summer' is that these insides are exteriorised, their private problems made public, as when the sun comes out, so too do the neighbours. For Hadley, when you see or hear your neighbours on the common, you realise how little it is you have in common with them.

> "The trouble is with summer, [is the] weather brings the neighbours out. Nobody knows anybody else, it's a place of people in transit: bedsits, students, housing association lets, first-time buyer flats.
>
> Even so, the people who live here get the urge to socialise outdoors in the summer just like everybody else. And then you lie awake until three or four in the morning, carried resentfully out to sea on a hostile swell of music and voices, the beat like chugging pistons from an engine room. Or you hear a flare of violence in the back lane, sudden and crazy as a dogfight. Or your neighbours open their windows wide to try to get a breeze going round indoors, and you're suddenly sharing in things about their lives you'd really rather not; their taste in music, their smelly cookery, their noisy love-making…" (Hadley, 2005: 91/2)

Hadley's female narrator in *The Trouble with Summer*, like the main protagonists in *The London Train*, is a liminal being in relation to these activities. Imagined as too old to join in, but too young not to feel jealous of the young people's activities, the narrator is lost in a suburban world in which she doesn't belong. Hadley implies that her character may have more in common with the past incarnations of the suburb than the present space she inhabits. For example, if we take a short walk away from Roath Lake we arrive at lines of smaller terrace houses, including Fanny Street. It is here that the narrator of *The Trouble with Summer* was tasked with clearing her dead grandmother's property:

> "When we were turning out the stuff from her house in Fanny Street we'd found an old black musty notebook with its covers hanging off, filled with tiny writing: so tiny you couldn't make it out without a magnifying glass and even then it was difficult because of the old-fashioned curly letters. It wasn't Gran's. The date written on the first page of the book was January 5th 1897…
>
> When we managed to decode bits of the writing, it wasn't very exciting. 'Thursday. Chops for Bert, egg for myself. Mrs. Hitchings called, says there's tableware at a good price come in at Collins. H. brought up the chair as promised. Milk of Magnesia…'
>
> …Sometimes I got muddled and imagined that the diary belonged to the old lady whose bedroom I lived in. I felt as if she and her furniture and her life in here were the solid presences and I was the ghost. ...Everything was the wrong way round: the lost past in the book felt solidly real and my living moment was a dream. Nothing specific had happened to me, to make me feel this awful nothingness." (Hadley, 2005:93/4)

Like Andrew Craig Williams (Chapter 3), this musty notebook was a material item that resonated with past lives, however mundane and everyday those lives may have been. It is this physical materiality (be in the form of specific items, rooms, or houses) that stores the ephemerality of being, as Halbwachs affirms:

> "since our impressions rush by, one after another, and leave nothing behind in our mind, we can understand how we can recapture the past only by understanding how it is, in effect, preserved by our physical surroundings. It is to space – the space we occupy, traverse, have continual access to, or can at any time reconstruct in thought or imagination that we must turn our attention." (cited in Rosenberg, 2011: 131)

As the narrator from *The Trouble with Summer* had also "fallen out of a very intricate piece of machinery which had been [their] life", the preserved events of the past offered a very real escape from her own rootless existence, and offered a compass point from which to direct a future.

> "The next day I started writing in my own words in my notebook, at the back. Nothing about unhappiness or being angry. Ordinary things. 'Got up. Bought tea bags, milk. Launderette tomorrow.' 'Decided to buy posters in town and put them up on walls.' 'Beans on toast. Yoghurt.' I was feeling better. I was holding on to a hope that I could build myself back together out of these solid things, make my life real." (Hadley, 2005:96)

This half-lost, half-found, betwix and between life resonates strongly with Cora's experiences in *The London Train*. Born and brought up in the faded splendour of Roath Park, in a house adjacent to Waterloo Gardens (see Map 5.1) her own ephemeral activities are stored in the physicality of the buildings and nature of the neighbourhood; as Hadley writes:

> "Cora had played in this park all through her childhood, felt as if it was yesterday the ooze of the stony brook between her toes, her mother's dread of broken glass and lockjaw." (Hadley, 2012:183)

Figure 5.4 Roath Park's stony brook. (Source: Author)

In some sense Cora is defined by Roath Park. This neighbourhood has been a constant part of her life, and due to the capacity of place to serve as an armature of memories, Cora's relations to these pasts and past-places remain very real to her. With the loss of her parents she has inherited the family house on the Lake, and seeks to renovate it, marking time and putting her own stamp on the place. To this end she knocks, "the two reception rooms into one, extends the kitchen into a new conservatory, sands the floors, paints everything white, [and gets] rid of most of the old furniture" (Hadley, 2012:170). Reflecting this process, geographer Carl Sauer (1925, 1963) postulated that places are 'palimpsests', that they can be understood as surfaces on which successive generations inscribe and remove 'traces' of their existence (see Anderson, 2010). For Cora however, her superimposition of conservatories, paint, and furniture did little to overwrite the material and mental traces that her parent's house had for her; despite her renovation the house was "still the same place as it had been when Alan and Rhian lived here" (Hadley, 2012:166). As Cora's life had not moved on from the setting of her childhood, she is stuck in the structure of her previous existence. In a similar vein to the narrator in *The Trouble with Summer*, the vibrant and durable nature of traces-past mock the emptiness of her contemporary situation:

> "Her mother had been agonised, wanting to pick up the dress out of the wet dirt, afraid to countermand her headstrong daughter. Cora had loved the weight of the skirts kicking against her limbs; she had loved the passers-by, dog-walkers in the park, stopping to watch; she had laughed at her mother.
>
> She thought of these scenes now with derision.
>
> ... She was living in her parents' house again, sleeping in her old room, although she had changed everything... [it] made her sick." (Hadley, 2012:166)

Cora thus no longer 'feels good' (de Certeau 1984:108) in this neighbourhood. In contrast to the memories of place that give many a topophilic association with locations (see Chapter 3), she has a topophobic sickness for the house and the lake. At this stage in her life she feels as if she should no longer be defined as a child of her parents or by her childhood memories, but as an independent woman who has reinvented herself and her place in the city. Despite her attempts to do so, the layers of memory remain both real and imagined in her everyday life.

Figure 5.5 Cora's Cathays: The Library. (Source: Author)

I now cycle the five minutes or so up the hill from Roath Lake, through the cemetery, to Cathays Library. It is here, or again in a fictionalised version of here, that Cora worked. From the outside the library is beautiful. Crafted in stone, its aged walls and roof have been recently and sympathetically restored. Inside, its huge, dark-timbered ceilings create a cathedral-like cavity, with amazing two-storey windows flanking the walls. I count the panes: 4 large groupings, 135 panes in each, all bordered by smaller glass margins. It has a wooden, school-assembly floor, without the smell of disinfectant and polish. Its interiors are more unfortunate – black book shelves, green carpet, pale green walls that blend into dark yellow. Municipal mixed with methodist. Here's how Hadley fictionalises this place:

> "The library was at a junction on a busy road carrying traffic in and out of the city from the valleys. It was a Carnegie endowment from the early twentieth century, built like an odd-shaped church with two naves at right angles and high windows of greenish glass, mournfully aloof from the squat, bustling shopping street of fast-food joints, quirky cafes, cheap mini-markets, hairdressers. ...The staffroom looked over the Victorian city cemetery, a conservation area for wildlife. Sometimes she ate her lunch in there." (Hadley, 2012:175)

> "[inside] there were encouraging panes of stained glass – blue and yellow squares with red diamonds – above the issue desk, in a strip around the base of a glass dome, where dead wasps collected in dingy heaps. No doubt the architect had had in mind a library as it might have existed in a Burne-Jones painting: dreaming members of the public opening their minds in a jewelled light to Tennyson and Keats, rather than to Large Print Family Sagas and True Crime." (Hadley, 2012:202)

Just like the graveyard, the library is quiet but used. An odd beep echoes as books are charged in and out. I sit at a white plastic table (garden furniture meets Ikea), all to myself in the corner surrounded by the shelves of 'family sagas'. I glance over the spines:

> Anne Baker: The Best of Fathers. Liverpool Love Story. Rebecca Tope: Fear in the Cotswolds. Maggie Bennett: The Unchanging Heart. Annie Murray: Where Earth meets Sky. Kitty Neale: A Cuckoo in Candle Lane. And over there – Tessa Hadley: The Master Bedroom.

Being in the library evokes the feeling captured by Hadley in *The Trouble with Summer*; each of the faces I see appear like blank canvases. These lives seem very different from the intimacy and shared experiences implied by Andrew Craig Williams (Chapter 3); everyone is 'busying' themselves (doing nothing?). Are these individuals more loose pieces looking for a jigsaw of which to be a part? It is easy to think of Mrs Hitchings from Fanny Street today being part of this place. Being here now.

Paul's Plot

Away from the the suburbs of Cardiff, Paul's family life is based to the north of the capital in the Monnow Valley, Monmouthshire. Unlike Cora, Paul doesn't have a past haunting him in the valley, but he nevertheless feels out of place in this area. In contextualising Paul's family location, Hadley again uses geography to give the reader an insight into how her character feels alien and other in the place in which he currently lives:

> "The weather was hot and fine. [Paul] went out with his friend Gerald, for one of their usual walks in the countryside. They followed the Monnow downstream; it hurried noisily over the lip of boulders and pebbles washed smooth, bulging under the thick lens of water. The path first hugged its bank, then meandered away from it across small fields with hedgerows dense with birdsong, bee-drone; blossom was snowed over the stumpy bitter blackthorns, the beeches' slim buds were fine tan leather, the still-bare ash dangled its dead keys. One of the great patriarchal beeches had come down across the path in a high wind only a few weeks before, its roots nakedly upreared, the buds at its far extremity still glistening with deluded life, a woodpecker's neat secret hole exposed at eye level, a raw crack in the wood of the massive trunk where it had hit the earth." (Hadley, 2012:22)

As the father of the family Paul is similar to the ‘patriarchal beech’ in the Monnow Valley, he feels uprooted and out of place in this location. As Cresswell (1996, 2004) tells us, this out of place-ness can have both geographical and social dimensions, and as an English writer in a Welsh valley, one whose economy is based primarily on farming, Paul sensed he did not fit in with his neighbours’ history or culture; as Hadley articulates:

> “He was friendly, but Paul felt Alun [like the other neighbours] always kept him at a deliberate distance... just because of what he would imagine was Paul’s type: English, opinionated, arrogant. ...Alun was small and broad-chested, handsome; he kept liquorice-coloured sheep on the hills and a small beef herd on the red soil in the better fields; they had a farm shop where his wife sold the fruit from their orchards.” (Hadley, 2012:63)

> “... [Alun] works the land; what are we? We’re nothing, we’re only playing here. This place where he earns his living is only our pleasure ground. That’s what he knows, he knows we feel it. If we live here all our lives, we can’t earn that out.” (Hadley, 2012:87)

Despite Paul’s lack of belonging in the Monnow Valley, Hadley does not locate Paul in a simplistic dichotomy between the local farmers (like Alun) on one hand, and the English, non-farming incomers on the other. The reader is told that Paul feels a kinship with the local history and culture of the valley, despite being aware he could not belong to it, whilst also feeling an antipathy to incomers who undermine this culture simply through their presence (as he himself is doing); as Hadley writes:

> “Paul lamented some of the renovations in the valley, ugly barn conversions for holiday lets. Cottages that were once the homes of agricultural labourers fetched stockbrokers’ prices now, as if the countryside was under some sick enchantment, in which the substance of things was invisibly replaced with only a simulacrum of itself.” (Hadley, 2012:23)

Paul is thus again positioned in a liminal space within these dichotomous positions of ‘incomer’ and ‘local’. He is not part of the landscape, and yet feels an affinity with it; he senses the changes to the valley are undermining something important, yet he is partly responsible for them, and he cannot share his concerns with his family or friends as they do not understand his point of view:

> "They all seemed to be angry with him, accusing him of nostalgia, of a regressive taste for everything old, of indifference to what had been unjust or caused suffering in the past.
>
> ...Paul couldn't summon the energy to explain that he had only meant the past was precious because it was different not better." (Hadley, 2012:40/41)

Paul is thus a liminal being between the positions of substance (local culture) on one hand and simulacra (incomer culture) on the other; he is unable to fully locate himself in either camp and is thus an exile from both. As a consequence, Paul suffers from an 'existential outsideness' in relation to both communities in the valley. Existential outsideness is explained by geographer Edward Relph through reference to a poem by Rainer Maria Rilke, as Relph outlines:

> "*'The new city was still to me as though denied and the unresponsive landscape spread its darkness as though I were not there. The nearest things did not bother to reveal themselves to me. The alley climbed to the street light. I saw how alien it was'.*
>
> Rilke's poem expresses a rejection of an individual by a place which he is condemned always to observe as though from outside. There is an awareness of meaning withheld and of the inability to participate in those meanings. This is the condition of existential outsideness that has fascinated so many nineteenth and twentieth century novelists and poets." (Relph, 1976:51)

Paul was thus rejected by the valley, and felt he could not play a part in its substance or spectacle. As Relph would put it, he was 'condemned always' to observe this place 'as though from outside'. Paul's existential outsideness is positioned in contrast to his best friend Gerald who resides in Cardiff, in an attic flat the reader can easily imagine to be located in Roath:

> "Sometimes Paul wished they lived in the city, and thought it was a mistake, their having chosen the countryside. Gerald's flat was at the top of a tall Victorian house beside one of the city parks. All the heat in the house rose up to his attic and beat in through the slates on the roof; his windows were wide open, but it was still stifling. While Gerald brewed tea, Paul stood at the window looking out into the shady spacious top of a copper beech, one in an avenue planted along the side of the park. A tinkers' lorry, on the lookout for scrap metal, cruised past in bottom gear, and a boy sang out 'Any old iron', riding standing up among the rusting fridges and cookers. Paul said it was the last of the old street-cries, resonant and poignant as a muezzin." (Hadley, 2012:58)

In Gerald's flat Paul felt a sense of home and belonging. Its location in the city, with its street cries and history, gave Paul a sense of substance he could share in and its positioning at the top of a 'tall Victorian house' gave Paul a sense of escape which echoed his own childhood dreams:

> "Paul could remember learning in a geography lesson at school about the layered living of the equatorial forest – his imagination had soared at the idea of animals that spent their entire lives in its canopy, never needing to come down to ground level. He had not wanted particularly to travel to the forest and see for himself; the knowledge that it existed was like a reserve in his spirit, a guarantee that spacious beauty existed somewhere." (Hadley, 2012:75)

Gerald's flat, at the top of a house in the canopy of copper beeches, could be seen by Paul to be reminiscent of the equatorial forest, somewhere in which his spirit could soar. Indeed Gerald's life was often appealing to Paul in other senses too, he viewed it as autonomous and undomesticated, a world where "he was able to keep his own times and lose himself in whatever labyrinths of reading or thought he strayed into" (Hadley, 2012:23-24). Although appealing, Paul was also fearful that if he actually had Gerald's life he would lose himself in it completely; Paul was aware he needed his family to root him, however uncomfortably, in the valley:

> "…Sometimes Paul thought that Gerald's freedom was what he wanted most and was deprived of, because of the distractions of his family. But he shrank from it too; what bound him to the children seemed to him life-saving. He thought of them as his blessing, counterbalancing the heady instability of a life lived in the mind." (Hadley, 2012:23)

Paul thus remained a liminal character. He was out of place in the city and out of place in the valley, he could no longer be a bachelor like Gerald, yet he was not quite a happy family man; he – like Cora – had no place where they could wholly belong.

Changing places, changing people: when plotlines move

Despite this lack of belonging, Paul was able to regain a sense of identity and purpose when he discovered that his daughter from his first marriage had become pregnant. On receipt of this knowledge he felt a strong desire to visit Pia in London and subsequently bring her 'home'.[1] On leaving the Monnow Valley and arriving in London, Paul faces up to the very different world that his daughter has chosen to live in,

> "Eventually he found his way to [her] block of council flats, bleakly unlovely, rising on a built-up island out of the torrents of traffic that roared unrelentingly all round it. It was hot, the petrol fumes rising off the vehicles was as thick in the air as a distorting glass, and it took him a few minutes to find the right system of crossings to use to arrive at the entrance. The block wasn't a tower, only three or four storeys, and around it there was a high wall, which made it look somehow, in its blue and white paint, like a scruffy container ship out at sea.
>
> ...It shook his usual idea of his oldest daughter to find her displaced here, when he had only ever known her insulated by her mother's care: in this place, despite the entry phone and the concierge, she might not be safe. Even while he was concerned, he was also interested by the idea that she must have chosen this, in preference to safety." (Hadley, 2012:46-47)

As humans are defined by person-place relations, as a consequence it is possible to create new relations and identities when we move to a new place. As de Botton (2002:147) explains: 'Our identities are to a greater or lesser extent malleable, we change according to whom—and sometimes what—we are with… move B to another environment and [their identity] will subtly shift in relation to a new interlocuter'". On arriving in London Paul realises that Pia was growing up, through her own pregnancy she was on the verge of changing generations, and through opting to live away from her family she had sought a new location in which to define herself. In a sense, therefore, Pia had decided to create a 'spatial division' in her identity (after Anderson, 2004b). She sought to articulate one sense of self (as a daughter) in the Monnow Valley, but a new version of herself (as a woman and Mother-to-be) in a new place (London). In following his daughter to London, Paul also found himself refreshed by his new environment: in

---

1 Although this 'quest' gave him a sense of purpose it characteristically kept him isolated from the broader world, as Hadley writes on two separate occasions:

> "He might be able to persuade her to pack a bag and leave with him there and then: he would take her home .... The idea of restoring her triumphantly made him emotional. His mission sealed him apart from the crowds around him." (Hadley, 2012:79)

this new place he found that his habits and routines had fallen away, and replaced with other defining features:

> "It was only when Paul had been in [Pia's] flat for several days that he took in that there were no books in it, none, apart from a tatty dictionary and a couple of recipe books. ... He had always had a superstitious fear of being shut up somewhere without books; now that it had happened he hadn't even consciously noticed. ...Staying over with Pia, he didn't care. He had brought something with him from [home] to read on the train, but hadn't opened it. Nor had he unzipped the bag with his laptop in it." (Hadley, 2012:98)

> "He walked among the crowds and down the side roads until he was tired, bought smuggled cigarettes from a street vendor, then stopped at a bar in Upper Street that had tables on the pavement and read the newspaper over a couple of beers. When he called Elise [his wife] she wouldn't pick up. He left a message, saying he was staying with his old friends Stella and John, he would be home in a few days. Stella was his BBC contact. The lie felt bland in his mouth, he shed it effortlessly." (Hadley, 2012:94)

Detached from his normal relations in Pia's fourth storey flat, Paul had become a different person. Here Paul no longer required his once-essential books, and could now wander the markets with no responsibilities. In this new version of himself Paul could also 'effortlessly' lie to his wife about what he was doing. Paul thus generated a new version of himself in London, for him identity was no longer a singular or uniform 'thing', but rather a fragmented, multiple, process (see Anderson, 2004b, see also Chapter 12). The new 'London-Paul' became an echo of his younger self, as Hadley explains:

> "Paul seemed to slip back inside that past time when he was heedless and twenty, as though all his substantial life between then and now melted away. Catching sight of his reflection once in a shop window… he was startled to see himself middle-aged." (Hadley, 2012:104)

Thus although Paul could redefine himself, however temporarily, in London, this redefinition was akin to him shedding the responsibilities and routines that had built up over his grown-up self. With these outer layers removed, he felt like he was 'heedless and twenty' again. However, despite calling forth this youthful version of himself, the inscriptions of time remained on his embodied palimpsest, reminding Paul of his middle aged reality.

Seeking out escape: the anonynmity of nonplace

We have seen how Tessa Hadley paints Cora and Paul as lost souls, umbilically connected to their places and family (respectively), but without a sense of true belonging. It is as if both characters inhabit a limbo or liminal state, out of place in their lives and increasingly ill at ease with the suburbs in which they reside. Hadley draws the reader's attention to this situation when Cora is stopped at Bristol Parkway railway station, on a journey from Cardiff to London:

> "It matched [Cora's] mood that Parkway was hardly a real place at all, hardly a building: bolted together out of steel at some point on a map, outside the city. Time wore away in the perfunctory waiting room, or stalking up and down the platform." (Hadley, 2012:300)

As Hadley notes, Bristol Parkway is 'hardly a real place at all', and it is possible to consider Cora and Paul's lack of belonging in the light of this type of place. Geographer Paul Augé defines such locations as 'non-places' (1995), which describe locations such as railways stations, motorway services, shopping malls or airports, modern spaces designed without individuality or soul, but which nevertheless fulfill a necessary function. As Putz outlines, in non-places individuals are reduced to "a mass of isolated individuals who engage in endeavors with little regard for others. [In these spaces they] do not form personal ties; instead, [they] engage in contractual relations" (2012:156). According to Augé, non places are "spaces where people cohabitate without living together … All the non-place asks of you is to plug in and pass through" (1995: 110). As a consequence of these characteristics, non places: "signal … the rise of the transaction over interaction, and the sad life of the lost citizen in the tragedy of contemporary civic life" (Sharma, 2009:129).

Although Paul and Cora do not actually reside in non-places, they nevertheless experience their suburban homes as if they were non-places in all but name. Neither feel as if they belong to the Monnow Valley or Roath, and they are reduced to performing hollow, superficial relations in these locations, rather than enjoying fulfilled and meaningful connections to them (see Arefi, 1999). Thus for Cora and Paul, their own homes have become non-places: just like Bristol Parkway, where human become simple 'passengers' who 'plug in and pass through', their homes have become places of being, rather than places of belonging. However, the anonymity of non-place means that it is also possible for these sites to become places of encounter, especially perhaps for non-people looking for somewhere to belong. And as it was with Cora and Paul, the place of the train offered the potential for lost people to find something more than a contractual relation to perform.

The London Train

Between England and Wales, city and country, Cora met Paul in the liminal non-place of the London Train. This is how Hadley narrates their initial encounter:

> "It was a few weeks since she'd run away from the fertility clinic, almost six months since her mother died. Her teaching had more or less finished for the summer, and she was throwing herself furiously into the transformation of the Cardiff house, telling Robert she wanted to do it up to sell it. ...The train wasn't full; she sat at a table by herself. Dark drops rolled sideways along the window glass. For no reason, her heart was beating thickly, as if she was expecting something, though she wasn't, she mustn't look forward, because there was nothing ahead, nothing.
>
> A man stopped beside her, carrying a cardboard cup of coffee from the buffet, a briefcase slung on a strap across his shoulder.
>
> - Do you mind if I sit here? I'm escaping from an idiot with a mobile phone.
>
> - How do you know I'm not one?
>
> He glanced at her, taking her in quickly. – You don't look like an idiot." (Hadley, 2012:229-30)

From this everyday contractual interaction, the liminal space of the London Train gave Cora and Paul the space in which to get to know one another. Although a non-place, its absence of expectations and responsibilities enabled a respite from their normal lives and a space that could be created of their own. Just through simple conversation – about their respective reading materials and other passengers – Cora and Paul joined their two isolated worlds together in the liminal space of the train. They bordered their territory on their own terms as they arrived at the next station.

"The train drew into the station at Swindon, new passengers got on, someone hesitated in the aisle at their table. They made no effort to move their bags from the seats beside them; both looked studiedly out at the platform, where those who wanted a different train seemed to wait in suspension, in a vague dusty light, cut off from the rain that poured in streams from the ends of the roofs. The person moved on: there were plenty of other places to sit. Neither acknowledged that anything had happened, but by the time the train started up again the atmosphere between them was altered, they were cut off together in their corner.

It turned out he had a house in the Welsh countryside somewhere – her geography was approximate..." (Hadley, 2012:233)

"...And do you have children?

He looked at her ring.

Something impelled her not to answer him 'not yet' or simply 'no'.

- I can't have them. We tried, but I can't.

- I'm sorry. Should I be sorry? Are you?

She shrugged. I would have liked it. But there it is.

- Might you think of adopting?

- No.

- OK.

It was a relief, to state the thing with such finality – as if she made it exist as an object to contemplate, stony, with clean lines and hard edges. With the loss of her parents behind her, and the loss of the babies she might have had ahead, she was withdrawn out of the past and future into this moment of herself, like a barren island, or a sealed box. It was easier to lay out this truth for the stranger's penetrating scrutiny, and not in expectation of any kindness. The hawk beak of his interest jabbed at her, as it had at the poetry book.

They could lose one another at Paddington." (Hadley, 2012:234)

Cora and Paul's island of intimacy was thus simply carved out of non-place by the arrangement of bags on empty seats and the non-engagement with other passengers at each station. However, the characteristics of non-place enabled their conversation. This meeting of strangers removed any expectation or judgement from their interaction, and they both knew they were free to jettison this nascent relationship as easily as moving carriages. In this freedom came a perspective and connection absent in their real life. This non-place had become a place in which they could both belong, with each other, at least for now.

In practice, their spatial bubble was easily burst. As Cora left her seat to use the restroom the world they had created had disappeared on her return.

> "When Cora had to get up to use the toilet she looked around dazedly at the other occupants of the carriage, suddenly returned inside herself and self-conscious about their conversation – how loud had they been? – as if instead of coffee they really had been drinking alcohol. Making her way back to him, balancing between the seat ends, after the unnerving swaying toilet and its complicated locking, she saw the cooling towers of Didcot power station float past the windows, the squat, fat pillows of their steam half-quenched in drizzle. The sight lifted her back into the surrounding routines of her life – they would be in London in less than an hour, she needed to shop before she got back to the flat – as if she was surfacing from somewhere underwater. When she reached their seats, he had picked up his book again and was reading as if he was reconciled with it. They had lost their momentum and sat in silence, restored to their separate existences." (Hadley, 2012:238)

However, the possibility offered by their short engagement in their own world of meaning meant that neither Cora nor Paul was wholly satisfied with the brevity of their encounter. As they arrived in Paddington and "the great station gave out its roaring exhalation" (Hadley, 2012:238), Cora realised all the things she didn't know about this man:

> "Dirty pigeons flapped like derision under the vast arch of the roof. Following, faced with his back, Cora was suddenly desperate at the idea of letting this unknown man go into the crowd where she would never find him again; she convinced herself that she might not be the whole of what she could be, without his knowing her. Hurrying behind, she willed him to turn round. And beyond the automatic barrier where they fed in their tickets, he did. He stopped in his tracks...
>
> - Oh, hello, he said. – It seems a shame, not to see you again. After we got on so well. Didn't we? When are you next in Cardiff?
>
> ...They didn't exchange phone numbers.... They were bound together, for the moment, only by the slenderest thread of an arrangement, an hour, a place (not her parents' house, but a cafe near the park)." (Hadley, 2012:240)

Looking out from the Terra Nova café the red pedalo now has an orange umbrella up. She is holding it, hooded against the misty rain, he is aimlessly spinning his legs. Not so difficult to imagine, among the mambo music, the flow of pensioners, infants, and general summer holiday (or what passes for it) flâneurs of the lake, that this would be the place for a liaison. Public, if not wholly front stage, the café is a veiled stall to the lake-(floor)show. When I asked Tessa Hadley whether the Terra Nova was the 'café near the park' where Cora met Paul she replied:

> "You think it should be café on Roath Lake but it isn't, that café is too municipal inside. In my imagination the ambience inside is like Coffee1 on Wellfield Road." (PC)

For me, Cora and Paul's café is a composite of both locations identified by Hadley. Located by the lake, but with the interior of a secluded, intimate hideaway it is at once an entanglement of geography and fiction alongside the real and imagined. However, it is also a liminal space – in-between the substance of Cora and Paul's relations on the train, and the simulated lives of their everyday. In this liminal entanglement lies the knot of Cora and Paul's subsequent relations. Does their relationship constitute a meaningful engagement between two people looking for a semblance of belonging in the 'tragedy of contemporary civic life'? Or is it better understood as two liminal beings simulating a relationship to escape the ennui of their existence? The possibility exists too, that their relationship is defined by a liminal entanglement of both positions.

Box 5.1. Cora and Paul entangled plotlines.

"When the day and the hour came round, Cora was almost too busy putting on white undercoat in the upstairs bathroom. By the time she got all round the window frame she was already late, and then when she changed out of her decorating clothes she realised that she smelled of white spirit and there was paint in her hair and under her nails. This seemed a doomed and desperate condition in which to seek out a love-affair. In the mirror she saw a caricature of herself, lips bloated, eyes bloodshot, charcoal eyeliner smudged. Fatalistically she almost changed her mind again and didn't go, only she couldn't bear the idea of the hours passing after she hadn't. Crossing the park under her resisting, buffeted umbrella, she felt the louring sky and sodden, thrashing trees were her own blemish, a weight she had to carry on her shoulders.

As soon as she came into the busy, noisy, steamy place, putting down the umbrella and shaking it out of the door behind her, he stood up from the table where he'd been waiting and came over and put his hands on her, holding her while she unbuttoned her wet mac, kissing her - right cheek, left - as if they knew each other well. Her mind was still in the chaos of wind and rain in the park. They were both breathing hard. There might be people in the cafe who recognised her, had known her parents: she didn't care. All the shops and cafes in Cardiff were poignant to her in that moment, suffused with a fond idea of home and the past.

- Oh, he said into her neck, - I thought you wouldn't come.

He smelled of cigarettes.

- I'm sorry I was late. I was painting.
- Pictures?,
- No, the bathroom window.

When they had got coffee and were sitting down he brought out a notebook and pen.....

Listen, he said, - I can't ever let you go again, English teacher, Henry James lover. Married to the senior civil servant. What's your name? What's your number?

...

- Cora. I made up names for you. But none of them was as good as that.

- This is ridiculous. We don't know one another,

- I don't want to know you. Not so that you're familiar, filled with familiarity, so that I forget the shock of you.

- You might not like me if you knew me.

- I like you. But that's the least of it.

What would she think of him, she wondered, if she was watching this from a cold distance, if she wasn't herself? Her nostrils tightened for a moment in disgust. She didn't want to be one of those women with hardened faces, joking about sex, lighting up at the idea of sex like an old, tired torch when the contacts are pressed, expert in techniques and devices.

But then, Paul - Paul, she savoured the name, as though she had always kept it ready, empty, as a mansion prepared for him - Paul wouldn't have wanted one of those women.

She thought: isn't it what I came for? Aren't I glad, that he is shameless? She didn't pull her knees away.

- Where can we go? Paul said." (Hadley, 2012:240-3)

Back in Terra Nova café, it is possible to argue that Paul, to some extent at least, has become a liminal being, a true inhabitant of non place. It appears that Paul does not want substantial attachment and belonging; his line: "I don't want to know you. Not so that you're familiar, filled with familiarity, so that I forget the shock of you", suggests he wants the thrill of the new without the commitment of the everyday. It appears he thrives on the "pleasure to be had in such abandonment" (Hadley, 2012:264), be it with Cora, in other affairs, or with his time-out from life in London. Perhaps it is only in these ephemeral encounters that he truly belongs. If Paul belongs in non-place, it appears that Cora is resigned to living there.

> "Once, Cora had believed that living built a cumulative bank of memories, thickening and deepening as time went on, shoring you

> against emptiness. She had used to treasure up relics from every phase of her life as it passed, as if they were holy. Now that seemed to her a falsely consoling model of experience. The present was always paramount, in a way that thrust you forward: empty, but also free. Whatever stories you told over to yourself and others, you were in truth exposed and naked in the present, a prow cleaving new waters; your past was insubstantial behind, it fell away, it grew into desuetude, its forms grew obsolete. The problem was, you were always still alive, until the end. You had to do something." (Hadley, 2012:311)

The reader is left, therefore, with Cora and Paul's lives of non-belonging and existential outsideness. Both are 'empty souls' which seek "a nice cosy little joint to be with his fellow man and not with the problems which haunt him. Not ever finding such a place, but pretending that it does exist. If not here then elsewhere" (Miller, 1947:XV in Relph 1976: 51).

Conclusion

In this chapter we have seen how Tessa Hadley examines places – in this case parks and suburbs – as not only the epicentre of dreams, but also places where you can attempt to sort out of your problems. It was in and around the parks of Roath that Cora and Paul attempted to tackle the ennui of their middle life existence by extending the liaison started on the London train. The motivation for this relationship was the absence of ties and connection both individuals felt in their home locations – their lack of belonging transforming their apparent hiraeths into abstract non-places. Both characters thus exhibited an 'existential outsideness' to their local geography, in the words of Relph, they portrayed:

> "a selfconscious and reflective uninvolvement, an alienation from people and places, a homelessness, a sense of the unreality of the world, and of not belonging. From such a perspective places cannot be significant centres of existence, but are at best backgrounds to activities that are without sense, mere chimeras, and at worst are voids." (1976:51)

We can see, therefore, how places are always influential on the human condition, but this influence is not always positive. Both Cora and Paul are out of sync with their respective surroundings, their home locale and culture do not fit harmoniously with their own. Their past relations, from their youth and family, appear to outweigh and overshadow their present person-place relations, and – unlike the narrator from *The Trouble with Summer* – they are unable to look to the minutiae of the everyday in order to relocate themselves. As a consequence, Paul looks to new person-place relations in order to refresh his sense of self; he finds a sense of home in this state of cursory and temporary involvement, in the spaces in-between the ties of family life and the void of isolation. Cora, in contrast, despite her attempts to overwrite her past, appears destined to be defined by her history; damningly perhaps, Paul simply describes her 'the park girl' in the end. Thus despite the possibility of place being 'mobile' and ever changing as new people, ideas, resources and materialities pass through them – for Cora the suburbs do not appear to change quickly enough for her to gain a new sense of place. In this scenario, it is often necessary for individuals themselves to move locations in order to place themselves and belong. In Chapter 6 we focus on a group of individuals who have done just that, from South Wales to mid Wales, and re-settled in Aberystwyth.

Part II

Kaleidoscopic Aberystwyth

Chapter 6
At the spinning extremes of existence: the thriving boiling seething places of Niall Griffiths

Spun like a centrifuge from the urban conurbations of London, Cardiff, and Liverpool, the characters in Niall Griffiths' novels whirl, drift and disturb the "remote, isolated, [and] difficult to reach" town of Aberystwyth (Griffiths, 2008:11). According to Griffiths in his contribution to Peter Finch's 'Real' series (see Chapter 1), Aberystwyth is "a place at the end of somewhere, a terminus" (ibid.). This is a town at the end of the line, folded in tightly between the mountains and the sea, and a place that is "equally inconvenient to everyone" (Fowler, in Griffiths, 2008:92). Despite this un(com)promising location, Aberystwyth remains a magnet for many seeking an escape from modern life. Niall Griffiths is one amongst this number; there is something in his 'salty blood' (ibid.) that is satiated by the torment of lives lived and lost in the coastal zone. As he admits:

> "...There's a strange draw to the place; get off the train and you can't go any further. Stand on the promenade and the next landmass is Ireland and, after that, America. Mountains pile up behind you and you feel that you can't go back into their bulwark mass but the sea, there, in front of you, is a blue world of possibility. And between that and the giant rocks is this small, intriguing town..." (Griffiths, 2008:11)

For those who have been attracted here, Aberystwyth's location becomes part of its charm. As a resident for three years in the early 2000s I was glad to brave the inconvenience of geography and relive a plotline through the town, its hinterland, and Griffiths' novels. I drive into Ceredigion on a grey spring day, with the hills covered in "that thin rain that just gets you drenched..." (Griffiths, Personal Communication (PC)). As I switch on my windscreen wipers and see the damp slate roofs and rocky outcrops around the town I am reminded how Aberystwyth offers a blank Palaeozoic geology (see below) from which individuals can assemble a life; as Griffiths puts it:

> "[Aberystwyth] is what you make of it; it supplies the materials, and what you build with them is entirely up to you. There are as many

'real' Aberystwyths as there are inhabitants of this peculiar, fascinating town. ...Look at the light with which that scene is lit; the bleached spill from the take-aways and the lilac lightning of the meat wagons and the ochre sodium of the streetlights and the pulsing rainbows from the clubs and bars. It's oneiric; it's fantastic. ...the town seems to lend itself to endless re-invention..." (Griffiths, 2008:12-14)

Figure 6.1 Aberystwyth from the hills to the south. (Source: Author)

Figure 6.2 Aberystwyth from the hills to the north. (Source: Author)

Griffith's Aber: The homeland of the drifting class

In this chapter we explore Griffiths' own re-invention of the town. As we will see, Griffiths' Aberystwyth does not shine a light on a town that is clean cut, middle class, and university-oriented (like the version presented by Collins (see Chapter 7)[1]; rather he looks beyond and focuses on the shadows this version casts (for an example, see Box 6.1). His is a town of drifters, drugs, and desolation, a place of splendour with hard edges, a place ringed by the hope of the sea and gothic wilds of the mountains beyond. This mix of brutish beauty is present in every description of Griffiths' Aber:

> "So this is it; town proper. ...You can hear gulls, and traffic, and crashing waves. Music from bedsits and pubs, outside the doors of which smokers lean and puff.... This is Aberystwyth. You can smell curry and fried chicken and exhaust and rotting weed on the nearby shore." (2008:80)

It is in this 'island' way out west that Griffiths' locates his novel *Grits*. The book explores the lives of those 'junkies an baddies an nutters' who inhabit the shadows of the town. These 'grits' are not indigenous to the area, but have been 'washed up' (Griffiths, 2000:18) by the centrifugal forces of modern life. Margaret introduces her fellow drifters in the following way:

> "A mean, take Squeaky Susie as an example; I asked her once what brought her to Aberystwyth an er answer was the kind uv thing that a lot uv people say around here:
>
> —Oh, y'know, oi just ferncied a day away, y'know, aht uv London. Came ere, got pissed, fergot ta get back onna bus. End av story..." (Griffiths, 2000:249)

> "We pass anavver train travellin in the opposite direction, tawards Aberystwyth ...No doubt bringin maw; maw expectant faces wawkin

[1] As Griffiths recounted in conversation, his version of Aberystwyth is not the one everyone from this middle class incarnation of the town recognises: "people would say 'this isn't the Wales I know' you know that kind of thing; people have their own aspects and own experiences and even in the Mayor of Aber, he said 'This [town] isn't the 'Grit?, this isn't the Aber I know', and I said, 'No, but it's the Aberystwyth I know', there's many different towns in there" (PC).

through tha station, blinking, hoping, lookin faw a certain somefing which they've neva bin able ta find any whir else." (Griffiths, 2000:481)

"... It's a town that evry drifter seems t'get to sooner or later. The homeland uv the drifting class, if that's not a contradiction an even if it is so fuckin what?" (Griffiths, 2000:250)

Box 6.1. Aberystwyth town

Map 6.1 The town of Aberystwyth

In his novel 'Stump', Griffiths' protagonists are caught in Aberystwyth's one way system, remarking upon the drifters and street drinkers they see in the author's plotline of the town:

"Rush hour in this small town and they join the farting crocodile of traffic inching its way round the start of the one-way system. Crawling past B-Wise and beyond that Argos and Iceland and Lidl and past the old train-station buildings now a Wetherspoons pub. At the roundabout by the twenty-four-hour garage this line of traffic splits into spokes and each one speeds up and Darren puts his foot down.

—Where to, Ally? Where djer reckon?

Alastair shrugs.—Dunno. Just follow the one-wayer, I suppose. See where it takes us, like.

—Aw shite. We're on the friggin prom." (Griffiths, 2003:196)

"One of the Aber halfwits, the wetbrains like, sittin on a bench by Phinikki's burger bar an shouting nonsense into space. So many of them in this small town, a disproportionate amount really shouters and screamers, street drinkers" (Ibid.: 87).

"Both of them look right then left, right along the curving sweep of the promenade, the hotel facades and frontages of halls of residence and Chinese take-aways and another huge hill at the end of it, this one with a funicular railway spine, and left at the large Old College the stone of its making the colour of bread pitted by spindrift and wind-borne salt. Ahead of them is the pier; pub and curry house and video outlet and flashing, bleeping arcade.

—Which way, lar?

They are blocking the road at the turning and a car honks behind them.

—Fuckin Mister Impatient there. Darren turns in his seat.—ALRIGHT, CUNT! KEEP YER FUCKIN WIG ON, WILL YER! Jesus fuckin Christ . . ." (Ibid.:197)

"Fuckin Aberystwyth, mun, mouth uv-a fuckin river like, seaside town by-a sea, come an visit an see a biggest fuckin concentreytion-a junkies an baddies an nutters at yew've ever fuckin seen, mun." (Griffiths, 2000:59)

Figure 6.3 North Parade, the end of the one way system. (Source: Author)

> "They drive back through the town wet and empty and climb the slope past the hospital and university back into the hinterland. Almost full night now and the street lights already sparse, the world shrinks to the misshapen splash of illuminated gleaming tarmac immediately beyond the bonnet and the needles of rain slanting through that leaning light. Wet blackness on each side and also above, the heavy cloud cover allowing no moon nor no dim cinders." (Griffiths, 2003:226)

In imagining an Aberystwyth that is the 'homeland uv the drifting class', Griffiths' writing does not glorify or voyeuristically exploit this subculture of alcohol and drugs. Rather he employs it to implicitly explore the factors which create such a class in the first instance. Griffiths writes his characters as more than stereotypes, giving them a history and a humanity which involves them hunting for a sense of belonging; as it states in the blurb to the book, their drifting is part of an "angry search for… meaning which they feel life lacks at the arse-end of this momentous century" (Griffiths, 2000). The reader discovers that the characters' drift from home was motivated precisely due to an absence of roots in their place of origin. For some of Griffiths' characters, like Roger, this was due to the lack of familial support and affection at home:

> "Hen wlad fy nhadau; an if so, en wha *is* this fuckin land? *My* father wuz a ole fat cunt oo ad me mam like a whipped fuckin dog, ulweys pissed up. an smackin us about like, fuckin laughin at me when a ceym ome from-a school speakin-a Welsh ad lirned. Oo once took us all on-a dey out t'Porthcawl an fuckin drove off an left us yer, left us t'find ar own wey back ome, nine fuckin yeers-a age a wuz. Teach yer t'survive, bollacks; fuckin teach yer ter lie an cheat an steal an ate yer own ole fuckin man. An is dad before im wuz-a seym, an probelly is an all like, so if iss *is* a land-a me fuckin fathers en it's a land uv old, fat, evil cunts oo cam keep eyr fuckin ands off-a fuckin botde ana scabby little dicks inside eyr fuckin kex." (Ibid.:90)

This lack of familial support in the 'Land of My Fathers' was compounded by the geographies of the sink estates in which characters such as Roger grew up. These neighbourhoods, like Gurnos in Merthyr Tydfil, South Wales, offered them little or nothing in terms of opportunities or prospects:

> "Graffiti on-a wall in-a New Gurnos like, biggest fuckin estate in-a ole uv Ewrop, fucked fuckin place like am fuckin tellin yew, Merthyr fuckin Titfeel, a fuckin *real* Wales fyer boy, none uv-a fuckin mountains or lakes yer jes fuckin sewage an dumps n boxes t'fuckin live in like, a *real* Wales, a place whir am fuckin *from,* Indyin fuckin reservation in-a middle-a Great fuckin Britain like (an Geraint the twat now standin on-a *real* fuckin reservation over in-a steyts like with feathers an spears an fuckin peyote) a pleyce-a gwehilion o boblach boy, too fuckin right, ewman fuckin weyst like, waste-a ewmans, gahbidge like, rubbish, shite, yer's fuckin graffiti on-a wall in-a New Gurnos: Croeso i uffern. Yew fuckin twat." (Ibid.:57)

Griffiths thus outlines a subculture within society that identifies no roots in the traditional sense, he identifies a drifting class who are disconnected and disassociated from their social and geographical communities. On a larger scale, Griffiths also outlines the ways in which the mainstream political and national cultures do little to offer a sense of collective belonging. In his view, modern culture has been reduced to little more than an interest in malign consumerism, undeserved celebrity, and reactionary versions of nationalism; as he elaborates in 'Max and Ronnie', a modern day reworking of the *Mabinogi* mythologies:

> "Max & Ronnie ... their country expects, they have been told, the millions of others of their blood and type are agog for them to prove that the heroic spirit of their ancient race is not dead and that their land can still spawn great emperors and kings and that the blood of knights and crusaders irrigates their limbs. ...[Their modern 'hero'] rides a big

> horse, a big noble horse, coloured ...with yellow on its legs and the man is wearing a shirt of red nylon with vertical yellow stripes and yellow piping at the cuffs and the yellow is as yellow as the For Sale signs lettering and the red is as red as newly spilled blood, that deep maroony kind of red, as red as a Manchester United home shirt for that is what it is, this rider having been born in Eastbourne and whose allegiances to football as to anything else were guided more by a perceived reflected glory than any kind of rooted loyalty." (2010:16/31)

> "The Jeremy Kyle Show is on television. Today's topic is Hooligan UK. And next up, ladies and gentlemen, we meet a man who's proud to call himself the 'Beast of Britain'. Wild applause and booing. A youngish man swaggers on stage. He's on camera. Regardless of why he is, he just is, and his face shows that he feels alive. This is where he deserves to be. In the spotlight. This is his entitlement. Number-one crop to his skull and a tight goatee with a tinge of red.
>
> - State of this bastard, Robert says. – No wonder he shaves his head. If his beard's anything to go by he's a fucking jinj.
>
> - Beast of Britain? Rhys snorts. – Bellend of Britain. State of him.
>
> Ah love mih countrih. Simple as that.
>
> So that means you should go abroad and smash up other countries?
>
> Ah love mih countrih. Me and mih merts, we love us countrih." (Ibid.:36)[2]

In the context of superficial celebrity and reactionary politics the reader is encouraged to frame the drifting of Griffiths' 'grits' as a political statement. Griffiths' characters wash up 'at the end of somewhere' in an attempt to escape from what they see as the vacuous culture of modern life. Finding no meaning or relevance in prime time television, neo-liberal democracy, and imaginations of nation (be it an imagined Britain or Wales, see B. Anderson, 2006), mainstream culture has failed to integrate these individuals into its hopes and dreams. As a consequence Griffiths' characters have drifted

---

[2] Griffiths talks of the research he undertook for this entry in the modern Mabinogi series with some distaste: "I spent about a year just immersing myself in reality TV and trying to read the biographies of people like Robbie Williams and Katie Price, oh god it was just grim. I drank a hell of a lot; it was just like people gleefully pounding themselves down into an insipid and colourless paste. Really fucking grim" (PC).

to the margins, both figuratively and literally. It is to this geographical and social periphery that the flotsam and jetsam of modern life have been washed up, creating Aberystwyth as an "anarchic centre [for] those who live apart from the mainstream..." (Griffiths, 2008:108).

In this sense, Griffiths is writing as part of a counterculture of radicalism that has always existed at the edge of the mainstream, and has written about in various ways through the novels of Irvine Welsh (1993), William Faulkner (1996), and Jack Kerouac (1999). In a culture that reduces coingredience with place to thoughtless patriotism or displaced football tribalism, Griffiths implies that his character's retreat to hedonistic thrill through drugs and alcohol is an understandable, if not necessarily laudable, step. In wanting something 'other' than the modern culture of the day, his characters have 'Othered' themselves from the mainstream, refusing their enclosure in prison-like 'reservations' (like the Gurnos in Merthyr) and seeking instead a different future. Rather than viewing these individuals as problems – as the pebble in the shoe of Aberystwyth, or the grit in its oyster – Griffiths' engages the reader to see that it is mainstream culture that is problematic through failing to identify the problems it creates or providing solutions to tackle them. As Griffiths outlines in conversation:

> "you know Leanne Wood, the Plaid Cymru lady, she said: 'I don't want my people to muddle, I want them to soar', which I know is a politician's sound-bite but I think she was on the right track, you know? Cameron [the (then) Prime Minister] was on the television last night saying, 'Well I'm for the British person that works hard, saves his pension, and supports his family', I mean for fucks sake, there's more to life than that you know?... Why not honour the miraculous quick of your existence and *do* something with it?" (PC)

Through focusing on the drifting subculture Griffiths bring a new perspective to it, his characters are not simply reduced to the 'Other' constructed by Middle England, to be reviled, ignored, or criminalised. For Griffiths they are not simply driftwood to be swept away, but instead he attempts to find them a place in this sweep of coastline way out west in Aberystwyth. As he states:

> "my characters are just kind of menial pips on the supermillennial screen you know, but I'm trying ... to give these small, what some people would see as miserable, lives, squalid lives, a kind of epic grandeur." (PC)

—Erm . . . what exacdy are we lookin for?
Ah shrug. —Nowt in p'ticular lahk. Just, ah dunno, some kind uv feelin ur summat a suppose." (Griffiths, 2000:170)

In line with Griffiths' mix of beastly beauty, despite or perhaps because of (ab)use of drugs and alcohol, his characters find a sense of value and belonging in Aberystwyth. With the hope of the sea in front of them and the immensity of the mountains behind, his characters become sensitive to the beauty and value of nature as a contrast to the grime and shallowness of modern culture. In line with Burke's description of the sublime (1761), his characters come to appreciate the world of nature. It is in the natural landscape that they find their meaning and sense of beauty. As Griffiths' Colm notes:

> "A walked inter town along the prom. Gulls an pigeons wer perched on the railings but thee all flew squawkin off when a coughed up another loader phlegm. Ug. Me cold's goin like, but the catarrh was still doggedly hangin on. A spatter big bundle uv green slimy stuff over the railings, into the sea. The sea looked fuckin incredible." (Griffiths, 2000:331)

> "A look ... down the coast ahead uv me. The last straggling street lights in Aberaeron and Newquay ar just about discernible through the swirls uv mist, miles awey like, shinin weakly out from an ungraspably huge line uv towerin mountains an cliffs. It looks beautiful, an if a had any fuckin breath left in me after last night's debaucheries then it would be taken awey by the view. This werld can be so fuckin gorgeous sometimes..." (Ibid.:324)

Finding a place from which they can appreciate the beauty of the world is more explicit for Griffiths' grits when we move out the town itself and into the mountainous hinterland. As I drive up and over Penglais Hill and "the world shrinks to the misshapen splash of ... rain slanting through th[e car's] light" (Griffith, 2003:226) his grits find some echo of themselves in the folds of this land, as Griffiths poetically writes:

> "They can be heard, sometimes, the mountains, they can be heard as they shift closer; a deep, low rumble from within the earth followed by a slow and persistent grinding, the hunger pangs of an immeasurable beast in subterranean slumber embossing its vast shape on the landscape, spurnose, mountainnipple, cliffrib, immense, recumbent and green. Why do they do this, the mountains, why do

they rush at the sea, why do they seem to seek that one element which always is and always will execrate and oppose them? Perhaps the possibility of union propels such gargantuan drift, the after-midnight keening of the Wicklow range at a pitch too high for human ear, the reuniting of a partnership once perfect, once powerful, now lost. So earth and water wrestle in deadlock, unstoppability against immovability, an eternal clash which will admit no stint or respite, and the breach between the two screams, knowing that like cilia in a digestive tract the mountains create a huge surface area in a narrowly-boundaried nation and that, when kind and whimsical, the ocean can on occasion permit escape. Concealment, then, or flight; the motives, aims and wherewithal will decide. You don't have to do anything. Just wait." (Griffiths, 2000:340)

Here Griffiths personifies the mountains and compares them to his grits, each rushing to clash with an experience which may result in a sense of union or destruction. In this way, he not only humanises the mountains and gives the reader a sense of a broader environmental sympathy with the forces of nature (a point we will return to below), but also territorialises his characters, giving them a landscape in which their lives are at once both a 'blip on the supermillennial screen' of geological time, but concomitantly something of epic significance and value. As Roger then Mairead put it:

"Fuckin amazin ese mountains ar, yer should fuckin see em . . . ey give off a weird energy, like, an it's stronger in some pleyces than in others, it kind uv fuckin throbs an hums, an when a reach-a spot whir a saw-astag (quickly, like, am speedin me fuckin feyce off by now, like) it's like a can feel iss. energy flowin threw me body, growin inside me like, sliker can feel me body bulgin an distortin with iss force an power ... a look down an a see me legs cheynge, see em an me arms go all thick an warped an bulgy ... a feel me feyce an it feels like-a fuckin lion's ead or summin . . . it's not scary, iss, in fact a trewth is it feels fuckin greyt. . . streynge tho; me mind pleys some odd fuckin geyms, it does, specially on-a acid, yer's so much fuckin shit bin threw it over-a yeers like . . . Fuck, tho, av never felt so fuckin *alive* like, so fuckin powerful mun, so fuckin *wild* . . ." (Griffiths, 2000:70)

"Sfuckin amerzin, this, out int forest, all green, trees sor thick an high . . . Trees aren't in regular, domed rows out here, or nor, theh mad, haphazard, an wild, followin nor pattern but theh orn,-which is just ow it should be, ah reckun. Ar feet crunch an squelch through t'leaves an bracken an mulch. Ah feel free here, rulleh free ...Totully free. Fuckin loveleh feelin this is. Rent arrears, fuel bills, council

> fuckin tax, theh can all go screw. This is t'plerce fer mih, oh yeh. If thuh wuh vodka flowin from uh spring, then, God, t'garden uv fuckin Eden this would be." (Ibid.:169)

We have seen how Griffiths understands the alienation and oppression felt by his characters in modern life of the city, and in contrast it is in nature that they become "exhilarated at the spinning extremes of existence" (Griffiths, in Brockway, 2004). By giving his characters an affective experience in nature, one that they cannot find anywhere else (through drugs or otherwise), he begins to root them and give them a place in the wild landscape. Here Griffiths' ideas resonate with a much broader tradition of nature writing that identifies the role of wild(er)ness in rooting the human condition. As Brooks explains:

> "When we talk about 'wilderness', what exactly do we mean? The word itself derives from Old English 'wildeor', wild beast. In ancient times it was a place hostile to man. The Bible equates it with 'desert', the last refuge for outcasts. ...Yet the significance of true wilderness ... is indeed a subtle concept, intuitively known years ago to Thoreau and Muir and a few others..." (Brooks, 1980: 253/255)

Henry David Thoreau himself explains the subtlety of the concept of wild(er)ness as follows:

> "I wish to speak a word for nature, for absolute freedom and wildness, as contrasted with a freedom and culture merely civil" (1862:1)

> "We need the tonic of wildness... we require that all things be mysterious and unexplorable, that land and sea be infinitely wild, unsurveyed and unfathomed by us because unfathomable. ... what I have been preparing to say is, that in Wildness is the preservation of the World. Life consists of wildness, The most alive is the wildest. Not yet subdued to man, its presence refreshes him...When I re-create myself, I seek the darkest wood, the thickest and most interminable and to the citizen, most dismal, swamp. I enter as a sacred place, a Sanctum sanctorum. There is a strength, the marrow, of nature. In short all good things are wild and free." (1894:238)

It is no coincidence that Griffiths' characters do not find the exhilaration of life in the forested plantations imposed and managed on many of the hills outside Aberystwyth, hills whose:

> "eternal contours are obscured by undifferentiated rows of Forestry Commission conifers. [This] manufactured uniformity ...is unpleasant and conceals like a cheap suit an immemorial world; ancient stone

> hills criss-crossed with the tracks and the stone remains of our ancestors." (Pryce, 2007:148)

In this acknowledgement of the power of the wild and immemorial, Griffiths echoes not just an anti-mainstream or anti-modern set of political values, but also a non-anthropocentric and pro-nature value system in the tradition of writers like Edward Abbey (1975, 1991). This non-anthropocentric, or ecocentric, position, as has at its root the science of *ecology*. The prefix 'eco' (from *oikos*, the greek for 'home'), inherently suggests that the planet is the native habitat for all living species, and suggests that all these species exist in a close 'web' of connection. Ecocentrism perceives that humanity itself is not the most valuable entity in the biosphere (i.e. it does not hold an anthropocentric vision of reality), rather it prioritises the process of ecological evolution itself. As Manes suggests,

> "Evolution means there is no basis for seeing humans as more advanced (or valuable) than any other species. Homo Sapiens is not the goal of evolution, for as near as we can tell evolution has no telos, it simply unfolds, life-form after life-form." (in Gottlieb, 1991:545)

Griffiths speaks of a sensitivity to this non-anthropocentric perspective as we walk into these immemorial hills outside Aberystwyth:

> "When I first moved to Aber ... we used to come up here and ...what struck me was the very first, my very first narrative that I read wasn't written by humans. They were written by animals or by the movement of migrating birds across the sky. That was when we saw the movement of the sandmartins up there [pointing up the hill] and we were all down here you know taking drugs and had a sound system and things and in the morning all the sandmartins were coming out and it struck me that when you're writing, you're actually becoming part of the natural world. You know you're making this mark you are creating a narrative out of your own experience, that's exactly what animals have done probably before humans had even evolved." (PC)

By adopting a perspective that is sensitive to the processes and practices of nature, Griffiths places his characters not simply in a landscape that is wild, but also that exists and lives in geological time. In this geographical and temporal location, they lose their preoccupation with the here and now and find meaningful roots in something broader – they have substituted the logic of politics and consumer culture for the logic of the land. This 'geo-logic' expands into an ongoing 'seam' adopted by Griffiths in *Grits*. He likens the geology of the area to the people who have become washed up in this land:

> "Grit: a wide, vague term for any sedimentary rock that looks or feels gritty ; . . [G]rits occur among the lower Palaeozoic 'greywackes' and are rocks that are on the whole fine-grained but contain larger, somewhat angular fragments (such as those in the Aberystwyth Grits formation) . ... Sometimes the term is used for any hard sandstone bed to distinguish it, by a conveniently short word, from other kinds of beds in a sequence.
>
> Wyatt, Antony, *Challinor's Dictionary of Geology,* Sixth Edition, University of Wales Press, Cardiff, 1986" (Griffiths, 2000:1)
>
> "From *A Personal Guide to West Wales,* anonymous authorship, privately printed, p. 3
>
> Fossiliferous, problematic, varied: geographically speaking, the Ystwyth valley area of west Wales is rich and diverse, so much so, indeed, that there are no better specimens in these islands than its Grits, its intrusive magma, its alternation of shaly agrillaceous mudstones and pale, hard sandstones. Aberystwyth town itself is sustained on such ancient, subterranean phenomena; mortared by ossified underground rivers of mud, it pays witness to the perennial battle of water and rock, the slashes of bright microgranite on the bared seacliffs the scabs and scan of this endless war." (Griffiths, 2000:2)

In this geology's 'fracturing', 'faulting' and 'contortions' (2000:49) we see a new way to perceive Griffiths' characters and a different timescale from which to view their vices and anxieties. We see how the 'grits' in this area "show a restlessness which will allow no permanence; these patterns, here, will never be static, nor complete, nor perfected, they will for ever be in a process of formation" (ibid.:49), and we are unsure if he is referring to his characters or the geology. In calling on such perspectives, Griffiths gives his grits an affinity and association with wild nature, weaving their connections into nature as strongly as society, giving them roots that stretch from Aberystwyth into the hills.

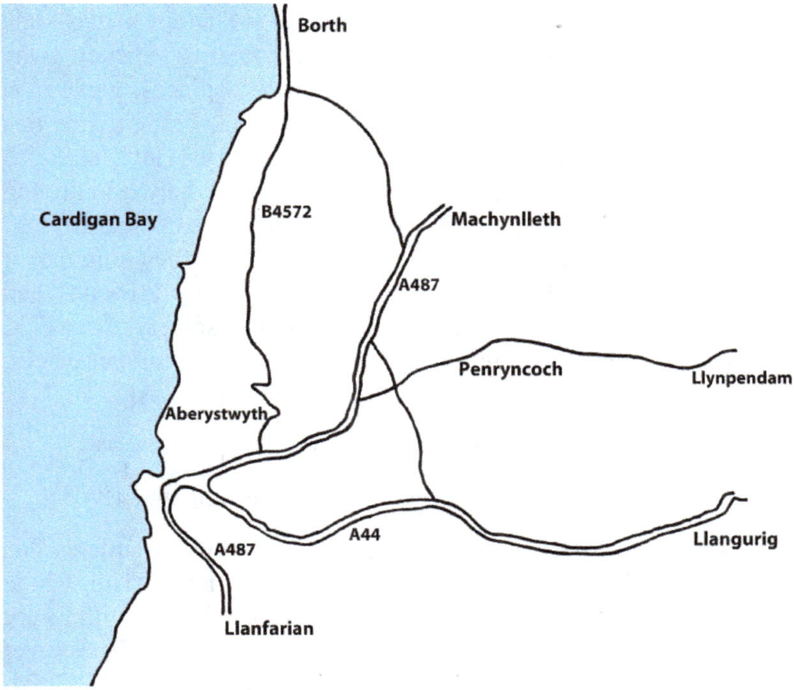

Map 6.2 Aberystwyth and its hinterland.

Figure 6.4 Into the wild hills above Aberystwyth. (Source: Author)

Such connections can be given further insight if we turn briefly to geographical and ecological writers. Doreen Massey, in her paper *Moving Mountains* (2006b), highlights how our perspective on contemporary society and nature is altered if we adopt a geo-logic timescale. Such a perspective, she argues, democratises agency away from the human, and acknowledges the agency of places, geological processes and other non-human elements. Warwick Fox extends this insight through his notion of transpersonal ecology (1995, see also Chapter 5). Fox suggests that humans' connection to the world around us is multi-layered in nature. This connection involves us 'identifying' not simply with ourselves as discrete individuals (be they singular or multiple, see Chapter 11), or solely with "our family ...friends and distance relations" (as we do in traditional modern society, after Fox, in Gottlieb, 1991:414), but also by our relations to non-human places, and even non-human species. As Fox explains, this process:

> "leads us to identify ourselves [not] just with our … our personal, biographical self… our family … [or] our community… [but leads us] to impartial identification with all particulars." (in Gottlieb, 1991:417)

Another proponent of ecocentrism, John Muir, further articulates this perspective in the following diary entry; he writes: "The whole wilderness is unity and interrelation, is alive and familiar... the very stones are talkative, sympathetic, brotherly.... No particle is ever wasted or worn out by externally flowing from use to use" (2004: 1911, entry 27 Aug 1869), and in the words of Griffiths and his grits:

> "These hills I know like I know the contours of my own body. They glow green today. So many different greens. Each leaf turns on its own tender hinge. We pass the Blaenplwyf transmitter, which allows us to watch *Sgorio* [a Welsh language football show, transmitted on S4c] as if through a blizzard." (Griffiths, 2008:29)

> "... Ah, what's the fuckin point, ey? That's why ah like stones, an rocks, an landscapes, cos the stones ur responsible fe evrythin, thur the furst cause, thur like God — evry time a see sum granite ah think 'daddy'. An the sparkly bits in granite remind me uv the fish-scales stuck te the skin on me da's hands an on is boots, an is hat… Fuckin incredible the wey evrythin's linked. Fuckin amazin wurld so ut is." (Griffiths, 2000:17)

Wild beasts and wilderness: from *Grits* to *Sheepshagger*

Traditionally, wilderness has to be located in a place distant and remote from human habitation. If we take this perspective seriously, wilderness can only exist in locations where humans are absent. Griffiths' focus on wild nature is therefore unusual as its location is so close to the civilisation of Aberystwyth – as he writes in Stump (see Box 6.1), it is just a handful of miles up "the slope past the hospital and university back into...wet blackness" (Griffiths, 2003:226). However, as Griffiths convincingly argues, although nature in the hinterlands is just a short distance from the town as the buzzards fly, it is quite a large distance in terms of altitude. As a consequence, wildness comes quickly in the darkness of the hills outside Aberystwyth:

> "A drive out past Penparcau an ontuh thuh Devil's Bridge road an just keep goin up, up, up into thuh mountains. It gets colder n colder the higher we go; so cold in fact that thiz even bits uv fuckin snow; at thuh sides uv thuh road. ...Thuh view up here's fuckin incredible. Yuh can see fuh miles, mountains an valleys, Cader Idris in thuh distance with mist on its peak like a fluffy hat.[3] It takes yuh fuckin breath away; so much fuckin *space.* A can see buzzards wheelin through thuh air in thuh valley below me, thuh size uv bloody wrens from ere. Never ceases tuh amaze me, lookin down on flying birds." (Griffiths, 2000:21)[4]

The hinterland of Aberystwyth is, for Griffiths, a true wilderness, as he explained to me on our ascent into it:

> "Even somewhere like this, you know, is fairly remote, you know you're not far from Aberystwyth but you now it's quite remote really. You still feel that. I mean from my house it only took us 10 minutes to drive up here but its pretty wild you know. ... I did something

---

3 In conversation, Griffiths acknowledges the poetic licence he has taken in suggesting you can see Cader Idris from the hills outside Aberystwyth: "Whilst 'Stump' is very geographically precise I like to mix these things up really and there's always poetic licence you know. In Sheepshagger when he climbs above Aber I say you can see Dolgellau, you can't you know but in that novel you can, narrative licence is a great thing" (PC).

4 Malcolm Pryce makes the same point in 'Don't Cry for Me, Aberystwyth' (2007, see Chapter 8): "The world here is old and misty, always misty, and it feels pagan. Somewhere beyond Ponterwyd you fall off the map into a desolate place designed for fugitives to hang out in and slowly starve; the sort of land where someone on the run would eventually get so spooked by the emptiness that he would turn himself in, even if it meant going to the chair; because the touch of the man who applies the electric conducting jelly to your temples is still a form of human contact" (2007:148).

> recently ...for Jarvis Cocker's *Night Thoughts* [a BBC Radio 4 series]. They sent me up the mountain at midnight with a microphone and we just recorded. We forgot we were in Wales it was like we were in Alaska, three feet of snow and howling blizzards. So I did my 50 minute thing and they said we absolutely love it but you're just too panicked. On the verge of hysteria. And I was. I mean there's not many wild places left in Britain but this is one of them and I think my writings always got to be aware of that." (PC)

It is in this wild place that we meet Ianto from the novel *Sheepshagger*. Unlike the drifting 'grits', Ianto is indigenous to this landscape. As Jordan outlines, Ianto is "a near-mute idiot savant with a mystical connection to nature who divides his time between roaming the mountains of his childhood and accepting whatever drink or drug is offered by his circle of half-teasing, half-accepting friends" (2001: no page). As Griffiths writes, Ianto is introduced as a wildeor character at home in the remote hills:

> "Dusk, summer, and Ianto leads Llyr and Llyr follows up the steep downrushing dingle, Llyr's cottage and the estuary behind and below them now afire in the sinking sun, a wet immensity washed in mild pink and harder orange intensifying in strata down the darkening sky to meet the eye-searing seething scarlet where the mudflat pencils a thick line against the sky, bisecting the far furnace of the setting sun. Llyr is panting, wheezing, occasionally coughing up phlegm shellfish which he spits out into the brittle ryegrass, where they cling to stalks like cuckoo-spit or odd land-molluscs unshelled. Ianto climbs steadily up the steep slope using his feet and long hands, prehensile and balanced like one unused to other forms of locomotion, say bipedal, upright.
>
> -Jesus, lanto mun . . . I'm buggered . . . how far now?
>
> Ianto shakes his head wordlessly and pulls himself onward. Upwards.
>
> This vast land they climb, its peaks and its plains like furniture designed by and for some titanic race was wind-stripped and rain-flayed of anything soft and yielding aeons earlier under a low young sun. There is nothing of any comfort here, all soil upholstery slashed and ripped and torn off and scattered horizonwards, tumbled over the world's edge in clods and slabs, if leaving just mud and protruberant close-packed stone in dense sharp ridges underfoot, squamous and scutellated like a reptile's back or the feet of a falcon. Ordovician country, named after the dark squat tribes it once sustained. Rock reinforced by lava flow and pumice to bulge and swell into mountain, quondam islands in the sea once here and now marked with the spiral and circle of petrified brachiopod and trilobite and conodont, languid graptolites drifting after plankton frozen in one instant of their

> incessant trawl and now caught stalking for ever. It is a land now as it was then, vast and rough and repellent and ragged; it must be scrambled up and through, upright motion is abhorred. You must dirty your knees and knuckles to let this land lift you up.
>
> —Fuck's sake, Ianto, mun . . . should install a fuckin escalator out yer aye . . ." (Griffiths, 2001:14)

Just as Griffiths' Aberystwyth is a very different place to the middle class provincial university town that the majority know (and is presented by Collins (see Chapter 7) and parodied in Pryce (see Chapter 8)), Ianto's hinterland is not the traditional 'green and pleasant' Wales. In an interview with 'Americymru', Griffiths explains the political importance of acknowledging that the wild is also present in the hills:

> "It's partly a reaction against the Enlightenment idea of Celtic peoples living lives of natural harmony and warmth; you know, 'let's not worry about these funny little people with their dancing and furry hats, they're all happy, they all link arms and sing going home from the mines to the hearth and a bowl of mam's cawl'. It's reductionist and self-serving and smug and undignified. Made by minds which can't see phthisis and poverty and self- and substance-abuse and loneliness and working twelve hours a day wresting spuds from rock only to be told on a Sunday that you'll be damned eternally for laziness. I chose the name Ianto [for the protagonist in *Sheepshagger*] partly as a nod towards the main returning character in Richard Llewellyn's *How Green Was My Valley* (well, less a nod and more of an abbreviated headbutt, really), which is symptomatic of this kind of Uncle Tom-ist nonsense. Stereotypes reduce, don't they? That's their job, to shrink in order to make certain people feel comfortable. They belong to prejudice, which is received hatred, and therefore a cliché of the most shrivelling kind. So, in *Sheepshagger* especially, I wanted to portray Wales as I know it; as an impossibly rich and wondrous and magical place which will fiercely fight back against any attenuation." (in Shaw, 2009:no page)

As the rain eases and we continue our walk through the hills, it is easy to find the raw landscape of this area beautiful. I say to Griffiths how it is also possible to identify the landscape's mixture of beauty and brutishness in his own writings (see Box 6.2). Griffiths replies:

> "Well really I'm saying this is a place of rot and shit and bone and faeces and all this kind of stuff and we have to accept it. Like buzzards, the buzzard's quite beautiful, a magnificent bird, kites are a magnificent bird, but their driving force is to eat, avoid being eaten, and to have sex, and that's basically it. They are beautiful as they are

> but you have to accept that the driving force behind them is to kill and eat you know? ... I read Desert Solitaire by Ed Abbey and what are the words he uses? 'It is ruthless and utterly without mercy but it's completely clean and beautiful' and he's entirely right. It becomes no less than a bloody model for living really and the development of your soul and I think that's how important it is and it's all linked in, the whole things linked in." (PC) (see Box 6.2)

Box 6.2. An avalanche of adjectives: land-scapes and literary-styles.

As we walk through the hills above Aberystwyth I am reminded of an interview Griffiths gave to 'Americymru' which talked about his written style:

Q: Your fiction often contains passages of highly charged poetic prose – this goes against prevailing notions that spare, pared down writing is somehow superior. How did you arrive at your prose style?

> "It fell on me, one wasted morning on Constitution Hill in Aberystwyth. I'd spent a few days in the mountains and at a lakeside rave and my blood and brains were bubbling, and words suited to expressing the madness came out of the sky. I was sick of minimalism; it can work very well, sometimes, but I felt then that the world was supercharged and that it needed a new expressive language. That kinda thing. You know we started this off talking about the reflection of the landscape in the sound of the welsh language and we've gotten to like the improvement of the soul and I think it's all linked in that way." (Griffiths in Brockway, 2004: no page)

Remembering this response prompted me to suggest to Griffiths that his written style is the literal equivalent of the wild but beautiful trees and rocks of the hills; he replied as follows:

> "Yeh, well I hope so, I hope so yeh, that's what I've been trying to do. I wanted to kind of run wild and have this notion that anything could happen even though the books' structures are kind of controlled, I want to have this notion that its running out of control, that its running out of the confines of the page and I quite like that idea, but also because I just want to bring language awake really you know and make it kind of resonant of mud and stone and everything else...

> I wanted it to go beyond onomatopoeia and evoke something more, [something] marrow deep. I wanted it to bring to mind the trickling of the streams or the sheep or sheepscapes or the dunes and craggy rocks, do you know what I mean? I wanted to write in a language that could only have come from this landscape really." (PC)

In extolling the mercilessness of wild nature and accepting its raw drive as a 'model for living', Griffiths' wild is not a pastoral idyll, but an awful sublime. As he notes:

> "it's kind of like finding a God – it's got everything to do with calm, and nothing to do with comfort. Rural Wales is a place of mud and death and shit and bone but it's also a place where connectedness is freely available and notions of re-birth declare themselves openly, and in that way, I find it immeasurably hopeful." (in Brockway, 2004)

Here we can see Griffiths adopting not simply a geo-logic but also an Abbey- or Fox-inspired 'eco-logic'. Indeed the value of the wild in forming a fully-realised set of person-place relations, as acknowledged by Thoreau (in particular the strength and marrow of nature found in a desolate swamp), is directly echoed in Griffiths' novel Sheepshagger:

> "I was trying to talk about [the value of the wild] and say how the landscape just absorbs into you and you absorb into it, you become part of it you know? Then Ianto ... lies in the bog and he feels himself melting into the bog this whole thing that you kind of get absorbed into the landscape and the landscape absorbs you." (PC)

> "He knows where he is, where he has fled to; the bog, the peat bog, in which he as a child would get mud-caked and sodden and play with toads and newts and lizards and even on occasion snakes. Confidant to the reptile he was, familiar of warts and forked-tongue-flicker, known to the slitherer and the slider and the lurker in quags. Still is: that property will never leave him, not until death itself swamps him and muds him and turns him slow to slime and slop, which is when such strange relations will be reinforced and nothing other or less.
>
> ... he continues to lie there anyway like some giant salamander, swamp-soft and bog-safe, his rushed breathing slowing. Smelling the methane leak and escape around and under him, seeing the stars blink weakly high above. Shooting stars and orbiting satellites. God how the night sky sometimes slides. Feeling the light rain on his skin cool his heated cheeks and dilute the acid in his guts.

> Ianto lying on his back in the bog. Snake-silent and lizard-still. And the moon has moved some distance across the soot sky before he hauls himself upright again, the back of him saturated and smelly and tentacled with hanging and dripping stems and stalks and leaves. Chickweed has bound itself around his legs and toadflax has infiltrated the laces of his trainers as if seeking a mate. He pulls a crumpled blue flower of bugloss from his drenched head and lets it fall back down to the mock ground.
>
> ...Of mountains, mud and mire is this young Ianto made. Fern-fronds his hair, stream-spume his drool. Night-time anthracite the pupils of his eyes. Snake-twined and standing to inherit this splitting earth." (Griffiths, 2001:25)

In Ianto, all of the Griffiths' themes are realised. His character has a primal connection to the landscape ('of mountains, mud and mire is this young Ianto made'), a (pre)historic sense of understanding and identification ('night time anthracite the pupils of his eyes'), and nature is seen not simply as sublime and convivial (in the view of writers such as Kropotkin, 1915, or Illych, 1973), but as 'red in tooth and claw' (after Tennyson, 1850; Ianto, for example, is 'snake-twined and standing to inherit this splitting earth'). This competitive element of wild nature becomes central to the narrative of Sheepshagger.[5] In Sheepshagger, Griffiths focuses on the native Ianto and his resistance to the incomers to (t)his wild landscape. These incomers do not acknowledge the land as primal site of connection, but rather reduce it to a simple playground for their leisure. Paralleling ecocentric writers

---

5 The competition and colonisation inherent to Griffiths' depiction of wild nature was also expounded in Grits:
"From A Personal Guide to West Wales, anonymous authorship, privately printed, p. 102

The unifying principle, the common squalid property of all forms of life-in-growth is competition, competition for light, air, nourishment, space to grow and feel and, primarily, move. Like vegetation, rocks too necessarily suffer the encroachment of aliens in weed- form and, like vegetation, need to banish or perish: thus wars, heroically sustained. Such clashes are fought out against a background of extinction, the loser's boon, a place grey as granite which crackles with static and endures for ever and for ever and for ever. So the unusual height of the coastline here is due only to the saw and slice of the sea; one element, one condition of matter performs a guerrilla raid on the other's magic arsenal and turns it to its own advantage. Such things do occur" (Griffiths, 2000:390)

like Edward Abbey (1971, 1991), Ianto resents such colonisation and is moved to violence to preserve the sanctity of his wild space. As the blurb to *Sheepshagger* states, Ianto, "pledges revenge not only on the English yuppies who have turned his grandmother's cottage into a weekenders' barbecue party but on all those who have violated him and the land that is his. This latest act of colonial oppression and desecration triggers his lurid and strange imagination into unspeakable savagery – embodying our most primal fears of physical threat, a world beyond our control". And as Griffiths writes:

> "The high moon pours cold down on Ianto's head, the insides of his ears wind-stung and wind-waters spilling from his eyes and rolling down his icy cheeks and he knows how it would be inside that kitchen, how warm and how safe, the log fire in the grate and the expensive whisky they're sharing and how it would be with him were he in there with the immense world-cupping knowledge comfort of living on and in his ancestral land his feel of the same soil his far forefathers dug in for their sustenance and ploughed and harrowed by hand, further back even to when barelegged ragged men with Ianto's same blue eyes laid traps, on rain-strafed mountain tops for boar and for beaver, fought bears and wolves and shaggy-tank aurochs with tree-spear and stone-axe, built stone houses beneath snowlines and waterfall and were the only knowers of this mad land. Which they can't feel, these new owners, ignorant of the particular preterite here, its knowledge and possible belonging, they can't feel that connection in their blood; although, which burns in lanto more, they behind the double-glazing in the fire's glow and the whisky; and their arm-touchy gestures and their head-back laughter don't seem at all perturbed or attenuated or even at all bothered by this lack. We want for nothing, this tableau behind glass and safe from wind and rain says everything we could ever need is right here with us. We want for nothing. We have it all.
>
> Ianto stands and stares. Fresh indulgence-pimples on his cheeks throb in the cold night wind and harden into callosities. Weak rain frost-dots his brow. There is a stone in his right; hand, petrified apple, which he raises over his shoulder as if to hurl, then lowers then raises again then lowers again." (Griffiths, 2001:22-23)

In *Sheepshagger*, Ianto becomes Griffiths' version of wilderness personified – he is the 'wildeor', the wild beast. Following Devall & Sessions (1985: 110) he is from the land and has developed a primal sense of place, he is not the conqueror of the land but the person fully in tune with his geological and ecological rhythms. And unlike a nature that is convivial and co-operative, he is ready to resist and violently oppose any colonisations. Ianto's actions could be understood as simply a form of anti-colonial, and anti-English, resistance. However, in line with Griffiths' Mabinogian critique of narrow-minded nationalism, and his sympathy with more eco-logics, it is better read as a challenge to a cultural trait that ignores the value of nature. As Griffiths states, the challenge Ianto offers is:

> "more class-based than nation-based, I feel, and I don't want to augment anyone's sense of indignant victimology (nor, indeed, dissolve it; we all need crutches, don't we?), but I do despair at the Playground Wales mentality that a lot of wealthy English people have. I despair at the smug and soul-less attitude that assumes that everything can be bought, that everyone has a price." (in Brockway, 2004:no page)

Ianto's stand is thus a primal response to the lack of appreciation for the notion that in wilderness may be – for both grits, drifters and us all – the preservation of the world.

Conclusion

We have seen in this chapter one author's reinvention of the town and hinterland of Aberystwyth. Through the lens of Niall Griffith's entanglement between page and place, Aberystwyth offers a destination for drifters, a draw for individuals who have nowhere else to go and no hiraeth to call home. These drifters not only bring wildness to Aberystwyth's streets, but also find in the wilderness of its hills a new form of belonging to that of modern life. In *Grits* and *Sheepshagger*, Griffiths offers us a view on how these new forms of meaning and connection can mobilise primal passions, mustering defence and loyalty, not to an imagined nation or modern notion of celebrity culture, but to a relation with the natural world that is pre-modern and perhaps pre-literate in nature. In Griffiths' plotlines the land, in all its brutishness and beauty, offers a model for living that draws attention to 'eco-logics' as an alternative mode of belonging. By placing ourselves not simply in towns but also in wild landscapes, Griffiths suggests that by paying closer attention to nature we can recognise "how the stones echo" (Griffiths, 2008:109) our own existence.

Chapter 7
One part memory and one part imagination: the entangled plots of Richard Collins

> "How's the novel?"
>
> "It's going very well. Actually too well."
>
> "How's that?"
>
> "Its getting itself mixed up with reality."
>
> I drink my tea and muse on the power of fiction. There's no escaping it." (Collins, 2004:122)

I have my suspicions that Richard Collins is a secret geographer. We had arranged to meet at his house in a village outside Aberystwyth on what happened to be a beautiful spring day. In his cottage-snug kitchen there are shelves filled with titles I'd recognise in a human geography library. We begin to talk of this book, and his books, and the power of geography and fiction. There really is no escaping them. After a rejuvenating coffee, we drive a short distance to a place he'd suggested would be perfect place to walk on a day like today. In line with his geographical studies, the location he kept a secret. I hoped we'd end up at a sarn, or glacial moraine, on the Cardigan Bay coast; a place where I'd imagined scenes taking place from his novel *The Land as Viewed from the Sea* (Collins, 2004). I was not disappointed.

In a way, Richard Collins' Whitbread nominated novel is the reason for this project. As a not-so-secret geographer, I have always been curious about the relations between people and place, how people are defined to some extent by the places around them, and how places become inveigled and intertwine with our identity, memories, and lives. *The Land as Viewed from the Sea*, along with Collins' other novels, *Overland* (2006) and *The Quality of Light* (2011), explore these themes. As Ali Smith explains, *The Land as...* is centred around the premise that:

> "A man is writing a novel and showing it as he does [so] to his friend, on whose farm he is working. The novel is about a man out at sea on a boat, considering a past love affair. 'None of this is real,' the writer

insists to the friend. But it doesn't take long for the real to affect the fictive and for the friends to become caught up in not just the composition and its sources, but also the unsayable things in their lives." (Smith, 2004:no page)

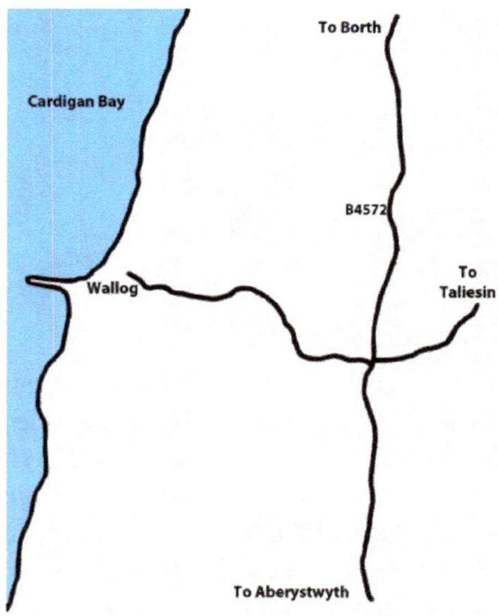

Map 7.1 Coastal Ceredigion, north of Aberystwyth.

As I parked the car and we began to walk down a track towards the sea, I ask Collins about the origins of the story.

> "The story came about because I did some sailing ...up right around the Llŷn peninsular and I looked across the bay at a place which had some important memories for me ... I love being on the shore looking at the sea but when you are out there it's really boring! Where those guys are now [pointing out across the bay at a boat on the water] I find that boring [but it did] give me an idea for a story." (Collins, Personal Communication (PC))

The story juxtaposes three constituent elements, the sea, the land, and the shore. As Collins describes, (all) at sea is a man on a sailboat, who is "looking back at the shore and reminiscing on a relationship..." (PC):

> "...Of course the sea element... stands for loneliness and it could sound like an arty literary device, a very 'clever' metaphor, but actually that guy on the boat would feel that the sea was a threatening,

> unpleasant place, especially if he associated it with his loneliness and associated the shore with his past love affair." (PC)

Collins fictionalises the affective experience of this element in the following way:

> "Now he has heaved to and knows that it will be alright. This is a safe way to spend the night on the water. He must only stay awake and watch the sails, tolerate the cold, the wet and his tiredness and wait for dawn. He had dreamt of finding some sort of peace out here, a meditative state, an empty mind. Instead he has been obsessively looking back at the land and thinking about two people on the shore." (Collins, 2004: 155)

On the land itself, the two friends (Julian and the narrator John), labour on a farm; according to Collins they are "very much rooted to the land, they're working on it with their hands" (PC). As the seasons change, John writes his novel based on his reminiscences of a love affair, and shares his writing with Julian. The shore is the location for the third element in the novel, the site of the love affair. As Collins states, "the two lovers are located at the meeting point of the land and sea all the time. You can hear the waves breaking in the background. It just worked really well that they're on the edge" (PC). For the couple the shoreline became the place where, "the[ir] soul[s were at] the centre of an elemental tug-of- war" (Collins, 2004:68); their tempestuous, chaotic affair changing John's life as he succumbed to a "powerlessness, [an] escape from reality (whatever that is), [an] insanity, [and] loss of control" (Collins, 2004:179). As Niall Griffiths reviews, the coming together of the land, sea and shore was crucial to the plotline of the story:

> "It's amazing, what the sea decides to jettison. From a boat, the town [of Aberystwyth] itself can look like jetsam, thrown up by the waves. Exploring it is simply beachcombing. Aber-based novelist Richard Collins knows this; his *The Land as Viewed from the Sea* is about precisely that – love, betrayal, creativity and destruction – everything that resides at the tide-line." (Griffiths, 2008:31)

Figure 7.1 The track that leads to the sea. (Source: Author)

Geography and controlled ambiguity

As we continue walking a narrow track down the valley towards the sea I tell Collins how, when I first read *The Land as Viewed from the Sea*, I didn't pay attention to the biography of its author (and thus know any details about where it may be set). Nevertheless I was intrigued by the fact that no actual place names were mentioned in the text. Furthermore, I was inspired by the book's title and cover. As an avid kayaker (see Anderson, 2014), I had paddled on the sea around Ceredigion and South Wales, and the cover photo reminded me of the land as viewed from the sea around both Aberystwyth and Porthcawl. Coupled to the idiosyncratic perspective captured by the book's title, the landscapes described in the book were, for me, so obviously from Ceredigion. We begin to discuss these themes as we approached the shore:

> *I say to Collins that his novels set a strong emphasis on evoking place and geography.*
>
> " ... Yes [they do], it's because I'm interested in places as much as I am in people. ...I'm not happy with novels that don't have a description of place; that don't have the setting. If two characters are

> walking down a track in the first page of a novel [like we are doing now], I want to know if the sun is shining or whether it's raining, what season it is. I want to know the sense of place, without that I'd put it down and pick up something else." (PC)

For Collins the writer and Collins the reader, the evocation of people-in-place is fundamental to plot:

> "My writing is about personal relationships, inner thoughts and [external] landscape ... if you really want to read some of the best writing about place ever you should read the first fifty pages of Lawrence's *Sons and Lovers*; it so brilliant. It's about a coal mining town and it's about a family and their most deep motivations, instincts almost, it's about their conversations and their family life and it's about place; it's set so strongly in place and there's a sense of the season and everything. There's also a strong sense of community, how people fit into a place through community as miners would, very much as a fishing community would be attached to the place through their actions. So places to me are as important as the characters. I wouldn't dream of starting a novel without knowing exactly where it was located. It's absolutely essential. It's part of the imagery, it's part of the plot, it's part of the tone of it and because of characters and interaction with their environment in their every move, it's part of the characterisation as well." (PC)

Despite places being crucial to Collins' plot, the author chooses not to name the places in which he sets his fiction (see Box 7.1). I ask him why he has made this decision:

> "For me it's better not to use literal names because you can have so much freedom with the plot and with the imagery if you don't.... You can mix them around so [for example] our hero sails along this coastline from Pwllheli down here [points across the valley to the sea and the north] on his own looking back, pretty miserable, looking back at the land as he sails; well he sails passed Aberystwyth [to the south] and ends up at this estuary [which is actually to the north]. So it's a mish mash. I wasn't constrained by literal geography, but it works for the plot, it works for imagery sometimes and in a way, in a funny way because the characters are so rooted in place, it works for characterisation to have that freedom, to not be tied to a literal place." (PC)

To evoke a place but not explicitly locate and name it has a curious effect on the reader. In the absence of place names, descriptions of locations act as teasers, suggesting to the reader that these sites may be wholly imaginary, but they also may have a real world counterpart. It is as if the author is inviting the reader into a secret, semi-fictional world, one that may co-exist with a 'real' one, but where that counterpart lies is known only to the author. Despite this, a reader is teased into solving this mystery.[1] Whilst reading it is possible for the reader to decipher 'clues' about the novel's real geography by recognising landmarks and recalling locations outlined in the fiction. The entanglement of page and place thus has the effect of seducing the reader, inviting them onto the page and imagining the fictional world for themselves. It is as if the author has sketched the outline of a real place, and the reader is asked to add colour and dimension to it with their own memories and imaginations. As with many themes in Collins' novel, this idea is echoed in the thoughts of the characters in his story, in this case in a conversation between the lovers John and Catrin:

> "Then they turn towards the sea.
>
> 'You know the way the bay curves round here?' she says. 'It's like a big amphitheatre. All the way around there are people looking in from beaches and cliff-tops.'
>
> 'But there's an empty stage,' he says, 'there's nothing out there.'
>
> 'Is that why we like it?' she says. 'We can people it with our own imaginings'." (Collins, 2004:16)

For this reader it is intriguing to feel that the power of the novel may come in part from a reader's ability to unpick the entanglements of Collins' semi-secret plot. I discuss with Collins my thoughts on the mystery of location in his novels; and it transpires it is an issue he has considered deeply:

> "I like to call it – I made this up – 'controlled ambiguity'; leaving the reader some work to do and leaving them space to project their own 'stuff' onto it. It shouldn't be a mess, but it should be ambiguous. There's something too literal about solid place names and writers before me like Hardy and Lawrence have made up names, but it gives a more dream like quality to the writing [to omit them]. This was really important for the book as it is about the kind of illusory quality of passionate love; it had to be dreamlike to work on that level. I think someone said in a review it was lifelike and dreamlike at the same time. I was delighted as that's what I was aiming for, so the absence of place names is a part of that." (PC)

---

1 The sense of being 'in' on a secret is compounded due to the relative remoteness of the area, and the associated lack of familiarity that most people would have with the locations chosen.

Box 7.1 The Land as Viewed from the Sea: locating a mysterious somewhere in West Wales

For those literate with the geographies of Ceredigion, it is possible to make educated guesses at to the location of the following scenes in Collins' novel. In this case, the written description actively evokes personal memories in order to involve the reader further in the landscape and story. For those without this geographical insight, it is nevertheless possible to imagine the world created by Collins, and engage themselves in the plotlines through their own flights of fancy.

> "He has been tacking close in to the wind and heading out into the bay. Now he turns the boat, the wind fills the other side of the sails and he runs in towards and along the coast, the wind on the starboard beam now, the boat keeling-over to port. Closer to the shore he can gradually see more of the largest town in the bay. Like most seaside towns it looks better viewed from the sea. You can't see the railway sidings or gasworks but instead the long terraces of fine Victorian houses on the front and above them large handsome buildings looking out over the sea. The library. The university. Near to the top of the hill, the hospital." (Collins, 2004:76) [This is Aberystwyth as viewed from the sea]

> "In fact Catherine[1] is out of hospital the next day and he takes her home. He is able to drop everything else and spend some time with her. She has a rented room by the sea in a characterful but tacky village up the coast a little way. The houses stretch along either side of a straight road with shingle and sand down to the sea on one side and an expanse of low-lying marshland on the other." (Collins, 2004:79) [This is a depiction of Borth]

---

1 Like the secrecy and controlled ambiguity determining the geographical locations in Collins' novel, his character John also chooses not to name the lovers in the recollections he writes. As we will see, however, Julian offers suggestions for the lovers' names, and chooses (to John's distaste) two names uncannily similar to the real life lovers (John and Catrin): Sean and Catherine.

> "Sean and Catherine are at the small village station waiting for his train. The single platform faces away from the village and the sea and so they find themselves looking inland, across the tracks to the marshes and the bleak hills beyond. They are the only people there.
>
> 'Busy today,' he says.
>
> She looks at him inquiringly.
>
> 'Three crows,' he says. 'Oh, and two sheep in the field over there. I don't know what your brother is going to make of this place. Straight from London to the edge of the known world. A bit of a culture shock'." (Collins, 2004:100) [Borth Railway station, at rush hour]

The effectiveness of controlled ambiguity throws light on the important relationship between author and audience, and the power of language to represent and perform the plot. As Thrift tells us, language functions as a communication device that is both 'representative' (i.e. 'standing in' for things, practices, and actions), and 'performative' (conveying more holistic aspects of emotions and empathy). Through these functions, language "set[s] up the intersubjective spaces of common actions" (1996:39). This intersubjective space makes it possible for an 'author' and their 'audience' to communicate and connect with one another.

However, there is always an ambiguity in this 'intersubjective space of common action', and it is this ambiguity that Collins seeks to exploit. Language can never wholly capture – or fully represent – the real world of experience, and it can never wholly communicate emotions and affect. Life overspills language, and words remain an imprecise tool to capture its complexity. It is inevitable, therefore, that in this space between author and reader there will be both clarity and ambiguity. If exploited correctly, this ambiguity can leave space for the reader to step into the text and populate it with their own encounters and experiences, creating a plot that is not simply the product of the author, but is also created in part by the imagination and experiences of the reader. The pages and places formed through these spaces are now essentially collaborative, forged through the meeting of author and audience. It is through this interpenetration of experiences, representations, and interpretations that we come to 'know' these plots. This is precisely the process that occurred for this reader in *The Land as Viewed from the Sea*. Imagining the text was set in the geographies of Borth and Aberystwyth, but never wholly certain of this fact, I was able to fill in the gaps between Collins' words with my own experiences and imagination. Due to the geographical nature of the human condition, not only did I feel as if these places were part of my life, reading these places in a fiction also interpenetrated my life into these stories, and vice versa. Through using controlled ambiguity, Collins – perhaps unwittingly – had indelibly connected his fiction into my version of Ceredigion.

"The trouble with you, Julian, is that you're too literal-minded."
"What do you mean by that?"
"You want it all wrapped up, straightforward, understood. Nothing is ever like that." (Collins, 2004:70)

As the track opens up in front us, I realise we have ended up at the sarn at Wallog. As Griffiths tells us,

> "jutting like a lapping tongue into the steel-grey sea [is] Sarn Gynfelyn... an impressive structure that reaches five miles or so into the sea although only the first few hundred feet are visible. 'Sarn' can be translated as 'causeway' [and] is thought to be a moraine feature dating from the last Ice Age." (Griffiths, 2008:64)

I first encountered this sarn on the water. Kayaking up from Aberystwyth at high water I saw a beautiful house on the shoreline, seemingly empty.[2] Gazing across to this building, paddles idle in my

---

2 This house also has a presence in Collins' plotline:

> "From the beach it is possible to see the whole long curve of the bay stretching away in either direction. To the south, headlands are visible in the distance, the further ones fading and insubstantial like clouds sitting on the horizon. To the north there are mountains close to the coast and a long peninsula, its hilltops visible above the horizon, like a series of islands. In fact the last seeming-island is an island for real, sometimes the beam of its lighthouse can be seen in the night across the bay. Above the beach, at the foot of a small valley that runs down to the sea, is a large mostly empty building, the old stable block for the grand house that once stood close by. It is now used for the storage of farm implements except for the flat upstairs above the stables. The building faces south, looking over the stream, to the other side of the valley. Only a small window on its gable end looks out

hand, I felt my speed accelerate. Startled and unnerved, as if electrocuted, I was at a loss as to the cause of my involuntary movement. Just like Catherine looking out across the sea in the following passage, I had no idea of the presence of this feature and its effect on the media beneath me:

> "...As they look out over what could be air or water or maybe some other even less substantial medium they can see below them one long straight line stretching out into the haze. She has no idea what it could be.
>
> "It's a sarn," he says. "There are two or three of them in the bay, long straight lines of boulders stretching out to sea, just under the water. I think they're roads to Atlantis but people say they're something to do with glaciers. They're dangerous to boats at low water, you can run aground miles from the shore. Or see just a rock jutting from the sea miles from anywhere. Seals rest on them sometimes. It's strange that this thing that's underwater stands out clearly when everything else is so obscure." (Collins, 2004:55)

Figure 7.2 The sarn at Wallog. (Source: Author)

Like Catherine, Richard Collins and I now sit on the shore. The tide is low and the sarn is now uncovered and stretches out to sea. On a day like today, with clean air, a light breeze, and a conversation that invites reminiscence, Wallog is a lovely place. For many it is also a

---

towards the sea. There are no other houses around and it is a bleak lonely place, cold and exposed, with none of the softening influence of the trees that grow further inland" (Collins, 2004:7).

place where reality and mythology blend like the headlands into the horizon. As Griffiths writes:

> "Once, the story goes, part of Cardigan Bay was a very fertile low-lying land surrounded by dykes and floodgates. On the night of a thanksgiving festival, which happened to be stormy, the guardian of the gates, Seithenyn, got drunk and fell asleep and forgot to close the gates and the wind drove the sea through to drown the land and all that lived on it. One of the area's main walls is said to be the causeway at Wallog.... On still nights you can hear the church bells tolling beneath the waves.... Terry Davies's book *Borth: A Seaborn Village*, is an indispensable reference work to the area. He explains the etymology of the strange word 'Wallog', how it may stem from old Welsh meaning 'unusual, quirky, or not quite right', and how 'the phenomenon of the causeway promoted Other-world possibilities'." (Griffiths, 2008:69)

In a similar way that 'Wallog' (as both a name and a place) offers unusual, Other-world possibilities, *The Land as Viewed from the Sea* deals with twin themes which serve to con-fuse and problematize the real and the imagined. Firstly, the novel explores how life is entangled by both geography and fiction, and secondly, it examines how memory, authorship and truth are all powerful but unreliable entities. As we have seen, the novels centres in part on the land-based relationship between Julian and John, and the reader comes to know of John's past love affair with Catrin through the story he writes. Julian also reads of this affair as John shares his drafts over the course of their labour. Through this process the lines between reality and fiction, and past and present, are thoroughly entangled. The seeds of reality always exist in John's fiction, but this fiction also plays a role in their lived reality. As the following passages imply, this entanglement is perhaps an inevitable – and necessary – aspect of the human condition as we seek to gain perspective and understanding of the events in our lives.

> "'Isn't it difficult disinterring old relationships?' [John is asked].
>
> 'It was a very long time ago. And I'm fictionalising it, changing it. That puts a distance between me and the reality of what happened'." (Collins, 2004:58)

> "When we have both got our breath back [Julian] wants to talk about my story.

> 'Has it ever been like that for you?' he asks.
>
> The rain falls outside, dripping off the points of leaves, filling puddles in the farmyard. The air has the smell of summer rain after sunshine.
>
> 'I don't know. You dig up the past, idealise it, make it into fiction. Lose it in the process.' I shrug my shoulders.
>
> [As a consequence] 'All this is fiction,' I remind him.
>
> 'But it isn't entirely fiction. We both know that. And you're destroying something as you unearth it. Do you remember I spoke about my brother who's an archaeologist? He says that when you dig up the past you destroy it. Doesn't that matter to you?'
>
> I can hear a hard bitter tone in my voice as I answer him. I don't like it. I know its masking sadness.
>
> 'No, I'm okay with that,' I say, 'It doesn't matter anymore'." (Collins, 2004:38)

Here Collins' characters interrogate the entanglement between reality, memory and fiction. It is suggested that the process of remembering and retelling has the effect of destroying the reality of the past, yet it is also implied that this past (even when it was in the present) was never a singular thing; it was always difficult to grasp, identify, and articulate. In this way Collins suggests that we constantly make memory myths from our lives, weaving connections between fiction and reality, page and place. In *The Land as Viewed from the Sea* this explicitly occurs as the characters re-experience their lives through personal memory, or recount and retell them through literal story-telling and everyday conversation. Collins identifies that this re-experiencing involves a re-invention, a form of archaeology that (un)covers, redacts, omits, and destroys aspects of experience, and re-places them with new or different versions. This acknowledgement – ironically given the metaphor of archaeology – adds a new layer of complication and interest to the reading of the story, especially for Richard Collins and I as we sit by Wallog, discussing his novel and our own plotlines that intersect in this place. "I'm glad it works", he says, "It's like you and I sitting here. I've become part of one of my stories" (PC).

Being caught up in a fiction can thus happen for both the author and the reader. This entangling process removes the 'fourth wall' between author and audience and reveals the conceit that any novel is a discrete, fully formed entity that can be passively consumed. In contrast, a novel is a creation that is edited, spliced, and composed from fragments of the author's (and readers') reality and imagination. The reader thus wo/anders in an entanglement of (back-)story of their own and the author's making. In *The Land as Viewed from the Sea* Julian and John are both similarly caught up in John's fiction as they read and contribute to writing the history of his affair; as the following excerpts illustrate:

> "Julian stops the little grey tractor in the shade under an oak tree. Later, when he's finished cutting summer cabbages and lettuces, he will load his harvested crops onto the metal box at the back of the tractor and take them up to the farmyard. He passes where I'm crouched over, weeding a long bed of carrots.
>
> 'Why don't you give your lovers names?' he says. 'I've never read a story where the characters don't have names.'
>
> 'Well, that's how it is.'
>
> ...'The woman could be...' he pauses and looks at me. 'What about Catherine?'
>
> 'No.'
>
> 'And the man is Sean. A nice name I believe.'
>
> 'I don't think so'." (Collins, 2004:70-1)

> "'You've put some of my words into your lovers' mouths,' Julian says, concerned.
>
> 'I did it on purpose. To show how its only a story.'
>
> Julian is standing naked with the spray of water hitting his shoulders and running down his back. He is tanned and muscular from working outdoors. Particles of soil are being washed down to his feet and onto the concrete path. I look away across the fields. After a while he turns off the shower.
>
> 'You make it seem true – that only she brings out those words in him.' He seems petulant, jealous." (Collins, 2004:17)

> "At lunchtime we take a trailer-load of carrots down to the farmyard and then sit in the sunshine to eat. It is sheltered here and warm enough at this time of day to take our shirts off, perhaps for the last time this year. When we have finished eating Julian goes indoors for a moment, returns with the pages I gave him this morning and begins to

> read. I lean back against the wall of the farmhouse and close my eyes to enjoy the feeling of the sun's heat on my skin. I have a different feeling about him reading my story now. These are the words that I wrote last night, they are still fresh in my mind, and he reads them sitting right here beside me. Something of the distance between writing and reading, between writer and reader, has diminished. I keep my eyes shut and try to enjoy the sun while he reads. Actually I'm with him in those pages...." (Collins, 2004:151)

Understanding how the reader, author, and characters can all get lost in the fiction suggests a myriad of ties entangling page and place: between author and reader, between real places and imaginations, and between real lives, memories and dreams. These ties serve to complicate the relations between humans and geography; the landscape no longer simply stores our ideas and memories for us in a neutral fashion, places now become active agents in holding, moulding, and creating lives.

In these more complex entanglements, crossed-lines and knottings between places and times are possible. I mention to Richard Collins that such knottings existed for me between my imagining of Wallog and Ceredigion, and the cover to his book which depicts a view from the sea to the shore of Porthcawl, South Wales. This choice of cover image adds further complexity to the folded and scrumpled nature of Collins' geographical plots; as he states:

> "it's quite interesting that Porthcawl [is on the cover], a place I've personally never been to. It's good that when you turn the page of the book you were taken to this coast which you knew. I love that, I love having a mental geography, having a world of other places inside my head which I can visit inadvertently or sometimes on purpose." (PC)

In Collins' third novel, *The Quality of Light*, this elision between mental geographies and real locations is also central to the story. Like 'The Land as…', *The Quality of Light* is set in an unnamed city, but to those literate of South West England, the location bears an uncanny resemblance to Bristol. Despite this, Collins admits the story is a mixture of mental and real places which make it more accurately a story of 'three cities' (Collins, 2011b). As Collins admits, "*The Quality of Light* is set in a place like Bristol with bits of New York and a few other cities thrown in for good measure" (PC); like Michael, a character in this novel, for Collins the entanglement between the real and the imagined is inevitable and absolute:

> "'OK', he says, 'I was once in America. It was a long time ago. I spent some time in California and then bought a car and tried to drive it right across the country, Jack Kerouac style. I didn't make it. The car packed up three-quarters of the way across and I had to catch the bus. I had spent a lot of time imagining that journey before I actually undertook it. And now, all these years later, I can remember different places in the mountains and in the desert and the mid-west and so forth. But there's no way of knowing which are the places I really visited and which are the ones I made up in my head'." (Collins, 2011:164)

As a result of these complex entanglements, places for Collins are "mindstuff", in other words, "one part memory [or reality] and one part imagination" (Collins, 2011b). When I ask Collins to expand on this idea he says:

> "I don't know how true it is for other people, but for me it is absolutely. I want to use the word redolent but I'm not sure that's the right word, anyway there are layers and layers of memory in place. When I go back to Bristol [one knotting in the plotline of *The Quality of Light*] it's just heavy with memories, same as this coastline. I don't know if everybody else attaches memory and imagination to place quite as strongly as I do? I guess they do." (PC) (see Box 7.2)

In his second novel, *Overland*, the power of landscape to hold memories is appreciated by his protagonist Oliver and his narrator Daniel. In the following excerpt we see Daniel remarking on the redolence of what appears to be an ordinary highway:

> "But this isn't just a stretch of road – for me it's marked by memory. I've been along here three times before and each time my journey had a different purpose and emotional charge. ... So as we go along the road now I see little landmarks along the way that bring back the feelings that were on my mind before. Oliver was like this when we were travelling together, the landscape full of ghosts for him. But for me it was fresh and new. That's what I want soon – to be in a new place, somewhere that isn't cluttered with emotional memories." (Collins, 2006:188)

This wish to have a 'new place' 'uncluttered with emotional memories' explains the need for John in *The Land as Viewed from the Sea* to seek refuge on the water.[3] In the popular imagination the sea is a different type of space when compared to terra firma. As Mack

---

3 This also explains the appeal of the liminal place of the train to Cora and Paul in Chapter 5.

identifies, the sea is often framed as, "the backdrop to the stage on which the real action is seen to take place – that is, the land" (2011:19). As a consequence, the sea is often considered to be an open space, a place of hope and freedom (see Griffiths, Chapter 6). Despite the appeal of these views, the sea can nevertheless hold memories (as my experience of kayaking to Wallog evidences) as well as offering a fresh perspective on more familiar elements. However in practice for John, the sea offered no respite from the 'mindstuff' of geography, as Collins writes:

> "Landscape and memory have always had a hold on him. He knows that particular places, or even distant views of such places, will open up memories. After their relationship ended he would avoid certain places that stirred unwanted emotions. Sometimes he would get caught out by an unexpected view. But gradually new events left their imprint on the land. The old memories were steadily covered by new, the landscape defused. So why had a journey by boat along this coastline reactivated all this old stuff? Away from the land he should be emotionless, serene, safe. But these new views of the land from the sea have triggered memories, unearthed feelings." (Collins, 2004:83)

Plotting a conclusion: love, belonging, and geography

We have seen in this chapter how the intersubjective space between writer and reader can be used to invite the reader to bring their own imaginations to the plot. Through controlled ambiguity and harnessing the power of 'mindstuff', Collins has also been able to capture that onieric quality of the land and sea around the hinterland of Aberystwyth (see Griffiths, 1998). As Collins writes, despite his unwillingness to formally map his stories:

> "my writing is rooted in this place, my writing's rooted in this place of course it is.... It's not a book about Wales in a political or social sense, but it is about the landscape. I originally thought of it as a British landscape, the west of Britain, but people have read the book and recognised this very valley so I have to admit it's more rooted in this specific place than I imagined it was." (PC)

As a consequence of rooting his writing in this way, in a similar way to Whitman (2005), Collins is able to 'invite the soul' of the reader to imagine and reminisce, about how the past effects the present, and about how we are formed by our past relationships. As the temperature falls and we walk back up the track into and out of Collins' story, I muse on the knowledge that my plotline of *The Land as Viewed from the Sea* is a very personal one, and may not be shared by others. My reality (itself composed by an individual entanglement of place, page and personal memory) is simply one in a plurality that are themselves constituted by an amalgam of ontological facts, controlled ambiguity, unreliable memories, and poetic imaginations. Although Richard Collins and I have shared our walk and conversation, neither of us really know what that experience was like for the other. The ambiguity in our entanglement therefore isolates us; as language re-presents, the original experience is lost between a sea of metaphor and land of simile. If taken seriously, no-one can ever fully know another's experiences, or another's thoughts, heart or soul. This realisation is isolating, as Daniel in *Overland* admits:

> "[Oliver is] sitting beside me, shrinking. He goes into himself and his memories. I suppose the place is very different to him without his family here. He sees their absence. But to me this place is shiny and new and I'm here in the moment, awake not dreaming, very much alive. I feel that Oliver and I are in the same place but at the same time in different places from each other, like we're in parallel universes. Is it always like this, people only seeming to be on the same planet as each other? I look at Oliver, hunched up, lost in thought. I'm sympathetic but also I'm fed up with him. I feel I want to spend some time with someone who shares my universe. I want it very much." (2006:39)

The role that mindstuff plays in isolating individuals reminds Daniel of the importance of being able to find 'someone who shares my universe'. Although geography is vital to our sense of self, for Daniel, and for Collins, it is ultimately "people that matter most of all" (2011b). In short, shared understanding between people is a crucial element in the experience of belonging. Despite the importance of this shared understanding, it nevertheless remains an elusive phenomenon. In *The Quality of Light*, for example, Collins' characters Daniel and Isabel are sceptical about the possibility of two people wholly sharing any of the complex mindstuff that constitutes our places; as Collins writes:

> "'You know the hill at the back of the village, the pointy one?' Daniel asks, for no obvious reason.
>
> 'Yes.'

> 'The pointy one?' he asks again.
>
> 'Yes. You asked and I answered,' she says, only mildly irritated.
>
> 'That's the thing,' he says. 'It's not pointy. If you look at it from the side, from the main road, it's just the end of a long ridge.' He smiles. He looks as if he has been proved right in some way.
>
> 'What are you saying?' she asks.
>
> 'Everything is different if you look at it from a different place.'
>
> 'Obviously.'
>
> 'I mean everything. Life is different.'
>
> 'Yes.'
>
> 'And you hope to some day find someone who sees things in the way you do – who stands in more or less the same place.'
>
> She thinks for a moment. Sighs. 'I don't think it's ever like that, Dan. Not in my experience. But sometimes you can understand how someone else sees things. That's as good as it gets, I think. Literature seems like that to me. You get to see someone else's view of the mountain. You see it clearly enough to understand. Or you think you do'." (Collins, 2011:112)

From Isabel's perspective, the best thing one can hope for in a relationship is to "see someone else's view of the mountain"; however, perhaps true belonging between people occurs when they share a common view. On the evidence of the fictionalised relationship between John (Sean) and Catrin (Catherine) in *The Land as Viewed from the Sea*, one could conclude that this mutual inhabitation of the complex mindstuff of place is possible, but only as a transient and passing phase. This mix of mindstuff, memory and imagination is deemed the only thing that is really important, far outweighing what is factually or ontologically true, and once it has passed, its absence echoes throughout our lives. Such emotional resonance means that deliberations over 'controlled ambiguity' or 'the intersubjective spaces of common actions' are far from abstract retreats into theory or philosophical detachment, but rather are an affirmation of the importance of imagination and re-visioning to our sense of belonging. From the evidence of Collins' plotlines, in order to fully belong humans require memory myths more than we need the cold hard vision of factual reality.

"Getting towards evening, the air still and the sun half-obscured by thin cloud. The sea appears to have its own light coming from below or within and glows with a metallic sheen.

'It looks otherworldly to me,' he says. 'I don't think you can take a picture of that, can you?'

There's a long pause during which her face clouds over with a frown.

'I hate it when you don't understand my art,' she says. 'It's the most important thing to me.'

'Perhaps I do understand a little. But because you take such care to get certain images I think you're trying to capture the real scene. And then you take the photographs and work on them and distort them with a computer or wrap them up in a painting until there's nothing much left of the original. So I wonder why you took such care in the first place. Or you produce a series that tell a story that never happened. Or paint something that seems unrelated to what you started out with.'

She watches him very carefully as he speaks.

'Then I realise what you're doing. That you capture a feeling, a sense of what was there for you. And I understand that that is more realistic in a way than any apparently accurate picture. I mean, that's all we have really. Our perceptions. Feelings.'

She is radiant, smiling as if he's making love to her.

'I think you really do understand a little. You wonderful man.'

Her eyes light up and she looks childishly happy. There is a strange deep curve to her lower lip as she smiles.

'Let's have some food and then a serious drink', she says." (Collins, 2002:27)

Chapter 8
Poetic refraction and stovepipe hats: the gumshoe mystery that is Malcolm Pryce's Aberystwyth

> "There was a bus from the village to Aber every morning, but not one back. That is so funny. I feel ... you can't capture that sort of quirkiness directly; you have to apprehend it obliquely, poetically." (Malcolm Pryce, Personal Communication (PC))

> "'I have come to the conclusion this is the same trick repeated every spring by the old courtesan Aberystwyth. She cakes on the all-concealing foundation, and stands at the back of the chorus line, where the shadows are deeper, hoping that her faded charms will last another season, while the leg-kicking strumpets at the front twirl petticoats that flash and blaze like fireworks in the hot footlights. Is it not so, dear Louie?'
>
> 'I've never heard Aberystwyth described like that before but it captures her perfectly'." (Pryce, 2009: 103)

If you have ever tried to get a double-vented, Italian-cut suit from the Mafioso tailors of Swansea, you will know the sort of difficulty you can have in trying to pin down Malcolm Pryce. Like the characters and ambience of Steinbeck's Cannery Row, trying "set [him] down alive" is "almost impossible" (1945:5). If we are to believe his biography – and for the sake of a more poetic life, why wouldn't we? – he has been, "an advertising copywriter" (where he "created campaigns for the famous Singapore Girl and also wrote tourist promotional advertising for the former headhunting tribes of Borneo"), "...a hotel washer-up, a BMW assembly line worker, the world's worst aluminium salesman, and ...a lotus-eater in Bangkok" (Pryce, 2001:247). Perhaps most outrageously of all, it was also claimed that Pryce was a librarian in Swansea (Allan, 2006), an allegation he later denied (Pryce, no date). Whilst taking "passage on a cargo ship bound for South America" Pryce "set to work writing *Aberystwyth Mon Amour,* the opening book in the Louie Knight series. The first draft was completed "somewhere off the coast of Guyana". (2001:247)

As a consequence of his near fairy-tale elusiveness, I discussed with Pryce the plotlines for his Aberystwyth detective novels in the virtual ether. This dark netherworld of internet communication was a world away from the Wolseleys, rock factories and whelk stalls of his Aberystwyth prom. Across the six books that recount the travails of private detective Louie Knight, Malcolm Pryce's Aberystwyth is perhaps the quintessential "place of mindstuff" (after Collins, see Chapter 7). Half imagination and half reality, over its series Pryce has conjured a parallel universe for the town, populated by sadistic games teachers, druidic cleaners, femme fatales in stovepipe hats, and a good-hearted gumshoe who seeks to "bring justice to birth" (Pryce, 2007:193) in the "bizarre refraction" (Forshaw, 2001:no page) and "convincingly surreal vision of the mid-Walean seaside town" (Owens, 2011:no page).

> "'[Once] you got a week's free bed and board [in Aberystwyth] if you bought a return ticket at Euston. Nowadays they can't even cut the grass between the tracks. You know what it is, don't you?' said Eeyore. 'It's like that movie with Charlton Heston about the apes. You know the scene at the end when he rides that horse along the beach and sees the torch of the Statue of Liberty projecting out of the sand? Suddenly he realises he's been on Planet Earth all along, after it has been taken over by apes. That's Aberystwyth'." (Pryce, 2007:107)

From *Aberystwyth Mon Amour* (2001), to *Last Tango in...* (2003), *The Unbearable Lightness of Being in...* (2005), *Don't Cry for Me...* (2007), *From Aberystwyth with Love* (2009), and *The Day Aberystwyth Stood Still* (2011), more than any other novels in *Page and Place*, Pryce has created a universe of characters, locations, atmospheres and philosophies that entangle with the real locations of Wales. Like laying a cloth over a table, Pryce layers fictional mythologies over the town, creating a "rich geographical layer hovering, so to speak, above the physically comprehensible world" (Piatti and Hurni, 2011:218). In a similar way to the reconstitution of the town following a biblical style flood in *Aberystwyth Mon Amour* (where the dams of nearby Nant y Moch reservoir were breached by a World War Two bomber), Pryce has (re)built the town as a hybrid of Raymond Chandler detective fiction, noir films, and black comedy.

> "Looking back, it was surprising how well the old place had stood up to the great flood. The waters had passed over Aberystwyth like a giant car-wash and picked it cleaner than an alley-cat does the bones of a kipper. But not much had actually been knocked down.... Here

> and there – like missing teeth — there were gaps in the rows of shops on Terrace Road and Great Darkgate Street. Little squares of rubble, filled with oily puddles, flapping polythene and broken dressers housing families of rats. And bounded on each side by the image familiar from the photos of the Blitz — sides of houses torn away to reveal the contents, floor by floor, like dolls' houses open to view… In the absence of canvases by Canaletto and engravings by Dürer we resorted instead to something more modest: a nationwide appeal for old holiday snaps and postcards of Aberystwyth. Predictably it produced its fair share of pictures of the Sphinx and the leaning tower of Pisa because, as anyone who's ever been stopped by a traffic cop knows, everyone's a comedian these days. But the steady stream of ash-trays, salt and pepper shakers, and souvenir barometers with views of the town were enough to get us started." (Pryce, 2003:13)

Pryce has, in a way, storied Aberystwyth afresh. In doing so, he has added literary 'storeys' to its geography. Perhaps he has fallen short of constructing Leaning Towers or Egyptian cats, but he has contributed Moulin clubs, ice cream stalls, and seaside donkeys to the layers of entanglement that now define the town. These extra storeys/stories are based on an identification of and basis in the 'real' Aberystwyth; as Pryce states:

> "a lot of people write to me to say they feel I have captured a distinctly recognizable essence of the town, one that they recognise." (PC, and see Box 8.1)

Yet he goes on,

> "But it's only semi-real isn't it? As various people have pointed out, my version is a parallel universe version of Aberystwyth which is superimposed on the real one. The two versions are not exactly congruent: my Aberystwyth incorporates recognisable aspects of the real world, such as for example the Prom, Castle & Pier. Other parts are pure invention." (PC)

Box 8.1 Pryce's Aberystwyth: gangster druids and supergrass ice-cream vendors

Pryce's Aberystwyth, although 'superimposed' over the real one, is irrevocably entangled for many into their version of the town, indelibly intertwined like pink letters through a holiday stick of rock candy. As the following readers evidence:

> "there are few clues to the reality of Aberystwyth's incredible underworld of gangster Druids, supergrass ice-cream vendors, tough and capable schoolgirl sidekicks, twisted genius schoolboys, Dietrich-esque chanteuses and evil mastermind ex-teachers that exist in Pryce's books. The heart-rending accounts of the Welsh veterans of the Patagonian conflicts and the disastrous consequences of the climactic ad-hoc Dambuster raid on the local reservoir bring a tear to the eye of the most sane reader. It's a damn shame the local council saw fit to hush it all up, I say well done for telling it how it really is, Malcolm!!" (Kevin Newton, Abergavenny)
>
> "Unfortunately these books are little known in the States. There are rumours that the author's book tour was cancelled following pressure by the Twm Sion Cati Anti-Defamation League. However, there is one guest bedroom in the Midwest where "Aberystwyth Mon Amour" features prominently. When friends read it, I earn new respect. "Are you really from Aberystwyth? Is it really like that?" I can't let them down gently. Of course it isn't really like that. Obviously, Mr. Pryce had to tone things down for publication or no one would have believed him." (Matthew, Indiana)
>
> "I was a student in Aber 77-81, ... I'm back there as soon as I open the page!" (Sue Mappledoram, Northants) (all cited in BBC, 2012)

The entanglement of Pryce's Aberystwyth through, in, and over the real version of the town has even led to readers contacting the author to point out inaccuracies in his depiction of the place. As Pryce states: "In Aberystwyth Mon Amour, for example, I referred to the mosaic of Father Time on the tower of the Old College, but it is in fact Archimedes" (PC). A more obvious concession to those who wish a direct rather than oblique version of geography in the fiction of Pryce was the relocation of Louie Knight's office in The Unbearable Lightness of Being…'. As 'Angela from Chepstow' pointed out:

> "I have just finished reading the third book by Malcolm Pryce. What brilliant reads. However one thing is bugging me. He makes several references to actual street names – Eastgate, Pier Street, Bridge Street etc – but he also mentions on several occasions a street called Canticle

> Street. Has there ever been or is there in Aberystwyth a street by this name?" (in BBC, 2012)

> "I have just had a reply from Malcolm Pryce himself. "Hi Angela. Yeah, Canticle Street. Sorry you couldn't find it – there are a number of errors & inaccuracies in the real Aberystwyth. I've written to the mayor but it's no use. I'd stick to the books if I were you. :)" (in BBC, 2012)

These blips in the real geographies of the town Pryce weaves into this fiction, as Louie Knight moves his office ho(l)me from the mythical Canticle Street, to the street with the Bakers on it in mid-town Aber:

> "'...Your office is in Canticle Street isn't it?'
>
> 'Used to be, we've moved to Stryd-y-Popty.'
>
> Today was the first official day of business in the new office and my feet hesitated for a second outside the old one but I forced them on up Canticle Street to take the unfamiliar left at the top into 22/1B Stryd-y-Popty." (Pryce, 2005:22)

Aber's past and present

Pryce's version of Aberystwyth plays with very British notions of the seaside town. Ice creams and 'kiss me quick' hats still exist in many coastal resorts, but they do so as memento mori's for bygone eras of holiday innocence, saucy postcards, and amusement arcades. Into this heady mix of reminiscence, Pryce adds the faded grandeur that is not uncommon in a Victorian seaside town, and the remoteness of Aberystwyth itself (as we have seen in Griffiths, Chapter 6), not only gives the town an end-of-the-pier curiosity, but also an end-of-the-line rarity. There aren't many places like Aberystwyth anymore. This recollection of resorts past is further intertwined with a sense that 'modern' life (even the modern life of Aberystwyth which as we will see later is perhaps not that up to date), has lost something in the inevitable treadmill of progress, as Louie states in *Don't Cry for Me Aberystwyth*:

> "And that was it, the essence of our malaise: our forefathers were entirely free of that despicable modern vice, facetiousness. I stood transfixed for a while, infected with Eeyore's melancholy. He had reached the Pier now, or what was left of it after the storm forty years

> ago: a dilapidated shed bathed in thin grey drizzle, and Matterhorned with seagull droppings; but still undisturbed by those hot, dry Saharan winds." (2007:107)

Figure 8.1 Matterhorned in seagull droppings: the pier at Aberystwyth. (Source: Author)

This recollection of resorts-past and a loss of innocent-present adds a pervading sense of nostalgia and hiraeth for this version of the town; the reader is invited to relive their own entangled imaginings of holidays gone by, and childhoods lived and lost. This Aberystwyth is a place without sat-navs, without wi-fi, and even without public telephones. Pryce plays with these ideas of childhood and simpler lives with the omnipotence of Louie Knight's old school teachers looming large as threats to our detective hero:

> "I took the Llanbadarn Road out towards the mountains of Pumlumon, along the course of the Rheidol for a while. And then cut south at Ponterwyd on the A4120 towards Ysbyty Cynfyn. A sign told me I was taking the Pont Ysbyty Cynfyn over the Nant Ysbyty Cynfyn and that was reassuring to know. Before too long, if my car didn't give out, I would be heading towards Ysbyty Ystwyth. The world was full of Ysbytys today and I wondered what it meant. Not knowing the answer in Lovespoon's classes would have resulted in the board-rubber exploding next to one's ear like flak. Ysbyty Ystwyth — the map gave it a black cross for a place of worship and a black box underneath meaning one with a tower rather than spire, minaret or dome. It also had a little symbol to say there was a public telephone.

> Compared to Ysbyty Cynfyn, which had none of these, it was Las Vegas." (2003: 211)

> "I think you know Herod Jenkins [the games teacher]? You cured him of his lost memory, but somehow a lot of people wish you hadn't. Right now they are holding out somewhere in the hills up by Nant-y-moch. They say there's a sacred place up there, something sacred to Herod. I thought you might know where they were, you being a special friend of Herod and all that. In fact, I understand you've made a map of his psyche. What does it say? 'Here be dragons'?" (2003:212)

When the trainline reaches Aberystwyth, the track ends. There were always said to be 'dragons' at the end of the line, beyond our known cartography, and perhaps the majority of the readership of Pryce's books will also not have an intimate knowledge of the town. This remoteness and unfamiliarity transposes a freedom to the fiction writer, which is perhaps enhanced by the onieric quality of the air which facilitates the imagination. Aberystwyth does tend to invite flights of fancy. As sometime inhabitants, my friends and I always used to imagine alternate worlds down quiet Aberystwyth streets[1] with sleeper agents masking as pensioners and champagne speakeasies hidden behind curtains in the milk bar. I asked Malcolm Pryce if there is something about being at the end of the line, that encourages – or even necessitates – such (dragon)-flights of fancy?

> "Quite possibly. Part of the town's character has to do with the journey you have to undertake to get there. The train journey is epic, but if you arrive by car over Penglais Hill, the view as the town rises into view is quite astonishing; like coming upon the lost valley of some fabled warrior race. It's impossible not to feel a little skip in the heart when you see that view. It's all downhill from there of course, ha ha just kidding." (PC)

> "We reached the crest of Penglais Hill and suddenly, as it always did, that familiar sight of Aberystwyth appeared in the valley below like a faithful dog, making the heart glad: skeins of smoke drifting across the slate roofs, the battered old pier and the pointy turrets of the old college, all set against the backdrop of a dove-grey sea." (Pryce, 2003:255)

---

1 Like the Mayor of Aberystwyth (see Chapter 6), I did not live in Niall Griffith's gritty version of the town.

Pryce's Aberystwyth successfully entangles the real and imagined through effectively juxtaposing unlikely elements into his plotlines. Alongside the innocence of a fading seaside resort (the 'old courtesan Aberystwyth' wouldn't stand for those 'contemporary interpretations' of fashion or culture 'like you get in Swansea' (Pryce, 2001:126)), Pryce juxtaposes the cinematic, American, sharp-edged thriller. In this sense Pryce explains that Aberystwyth:

> "perfectly suited the purposes of my concept, which was to import a paradigmatic hardboiled dime thriller private eye in the Chandler mould into small town Wales. One reason for this is Aberystwyth has a name to conjure with – up there with Troy, Gilgamesh and Timbuktu. It never ceases to amaze me how many people have heard of Aberystwyth, far more than is decent for such a small town. I have even met former headhunters in Borneo who have heard of Aberystwyth. But of course, the truth is, I didn't choose Aberystwyth, it chose me." (PC)

Pryce therefore combines the innocence of the past and overlays a Chandleresque genre to unmask the "moral grime of the Aberystwyth underworld" (Pryce, 2001:171). This underworld was often left unreported by the local and national media (in both the real and imagined Aberystwyth):

> "The most underworked man in town was Meirion, the crime reporter on the *Gazette,* he worked fewer hours in a year than Father Christmas. Not because of a lack of material. There was enough going on to keep an entire department on overtime, but the money that owned the newspaper also owned the seafront hotels and the ghost train and the putting green and various other bits of tourist infrastructure. To read the *Gazette* you'd think we were a town full of Tibetan monks." (Pryce, 2001:37)

> "No one ever got rich fighting crime, just ask Eeyore — a cop for thirty years before he became the donkey man [on the beach]; more collars than a Chinese laundry, and he ended up just as poor. Aberystwyth, queen among towns for irony; perhaps nothing was more so than this: that there was so little money to be made in law enforcement when there were so many villains, so many laws to be enforced." (2007:50)

This entanglement of genre and geography has led Damian Walford Davies to describe Pryce as 'transvaluing' Aberystwyth (2012:208). According to Walford Davies, Aberystwyth has become "parochialized and globalized" through these novels (2012:8), interpreting them at once as a local archetype of the Victorian seaside resort, and transformed through genre into an internationally recognised trope of fiction. As a consequence, the plot of Pryce's Aberystwyth has become at once a peculiarly British, American, and Welsh hybrid, as the following excerpts imply:

> "She leaned closer and whispered, 'I haven't got to the best bit yet.'
>
> I forced another yawn. 'Don't tell me, let me guess: some man in a long back coat turned up asking questions about him.'
>
> 'Yes,' she hissed. 'But the point is, what did he want to know?'
>
> I shrugged.
>
> 'The valise! He wanted to know what had happened to the valise.'
>
> 'What's a valise?'
>
> 'A case, you idiot!'
>
> 'Why didn't you say that?'
>
> 'I thought you were a private dick, that's how they speak.'
>
> 'Where?'
>
> 'In LA.'
>
> 'This is Aberystwyth.'
>
> She snorted. 'Fat lot of good you are.'" (Pryce, 2003:46)

> "The train picked up speed and glided soundlessly through the wide watery silence. Condensation dripped in icy streams down the inside of the window and, outside, the world seemed to be taking a lie-in. The sun glimmered through the lemon mist above the estuary like a nightwatchman's lantern. Patches of water in the peat glistened and looked as if they had been cut out of the sodden turf with a giant pastry cutter. To the west beyond the dunes the sea was silent. At times like this the sight of the estuary in all its beauty made the heart gasp and long to turn back, a last final coquettish trick from that old whore Aberystwyth. Like a lover who catches you with a packed bag, tiptoeing down the stairs before sunrise, and calls to you from the landing, looking like she used to all those years ago when you first met at the Borth Carnival dance…. Bert pressed his face against the cold glass, straining his eyes to make out features in the soft misty world. 'Never thought I'd finally do it,' he said. 'Never, thought I'd leave like this, never in a million years.

> 'Get used to it, pal,' I said, already tired of his moping. 'It's Aberystwyth not Monte Carlo'." (Pryce, 2003:23)

In this self-mocking black humour there remains the very Welsh hint of hiraeth, the jaded heart of broken dreams, the connection to times past, and an affection for landscape and community (which I will discuss in more detail below). Pryce sums up this combination in the following passage which emphasises the importance of the peculiarly British flavour in this hybrid entanglement. Musing over the accepted truths of the thriller genre which mean that private detectives should always drink whisky, Knight deliberates why this may be the case, and in so doing, emphasises why a British location is essential for the setting of this boundary defying series:

> "I put the notepad down and took another drink. Calamity, who thought all private detectives should drink whisky, once asked me what was the difference between [whisky and rum]. I thought at the time it had to do with the taste, but I was wrong. People are wrong about everything. What is the difference? They both taste fiery and get you drunk, they both look the same and cost the same. But one is the distilled essence of cold, wet, miserable Scottish highlands. And one is the succulent ichor made from sugar and the distilled sunshine of far-off places." (Pryce, 2001: 215-216)

Box 8.2. Fate, ennui and Aberystwyth: "The spiritual befuddlement that dogs the man of Aberystwyth at every turn is thus an unavoidable part of his fate." (Pryce, 2009:103)

Philosophically musing over the differences between rum and whisky highlight how 'spiritual befuddlement' is never far away from those locked into the plotlines of Pryce's Aberystwyth. As the following excerpts illustrate:

> "At the north end of the Prom, beneath the shadow of Constitution Hill, Sospan was leaning on the counter of his ice-cream kiosk, squinting as he stared out to sea, as if the answer to the mysteries of life were encoded in the hieroglyphical waves." (Pryce, 2001:235)

> "I walked up the Prom in search of Raspiwtin and found him playing crazy golf. Of all the rituals of the seaside holiday it must be the emptiest. It isn't crazy; not really. Despite the discordant primary colours painted on the concrete, it isn't zany or madcap or subversive or anarchic; it doesn't encroach upon the line separating genius and madness. It is simply dull. The grass is made of cement, which gives no purchase to the ball, and therefore it is impossible to aim with any precision. The ritual survives for one reason only: in our hearts we notice a subtle resonance with our own fates. We too careen around a concrete rink for a while, ping from side to side across a garishly painted world the colours of which betoken fake joys, driven by insane forces, subject to incomprehensible laws and rules in which annotations and signatures next to meaningless numbers and letters." (Pryce, 2011:97)

Pryce's Aberystwyth thus plays with notions of genre, nostalgia, and hiraeth, but nevertheless produces a clear geographical plotline. Despite this clear geographical signature, the era in which it is set remains ambiguous. It is easy to suggest it is located in the past – Louie Knight drives a Wolseley Hornet (a saloon car made in the 1930s by a long defunct British manufacturer), and Calamity is paid 50p per day for her initial detective duties (at the time of writing 50p may just about buy you a pint of milk in Wales). However to be more precise about the era in which Pryce's novels are set proves difficult. In a similar way to the 'controlled ambiguity' created by Collins (see Chapter 7) which enables the reader to imagine their own geographies, the ambiguity of Pryce's historical location means Knight's Aberystwyth remains a timeless place somewhen in the twentieth century. I ask Malcolm Pryce if this ambiguity is intentional:

> "A lot of these things emerge or happen without conscious deliberation. But essentially the world is set in that neverland that fiction about superheroes occupies. When, for example, is Superman set? Or Batman? I know the Batman comics were probably written in the 30s but the world exists in a time of its own. My Aberystwyth is notionally set in the 80s but in many respects belongs in the 50s, a world of Enid Blyton type criminality, and no mobile phones." (PC)

The innocence of past times is summed up in *Aberystwyth Mon Amour*:

> "I headed for the dunes that edged the park; there was something eternally beautiful and reassuring about them, the sharp spiky marram grass that stung our knees as children looked soft now, like fur ruffled by the fondling breeze." (2001:172)

Whilst Peter Finch summarises the qualities of times past that effuse Pryce's novels:

> "Malcolm's novels ...seem to recall a lost moment; their protagonist, Louie Knight, seems a product of a sadly gone age, more easily described as 'innocent' than our own, when a basic decency had a say in human affairs and dignity was something to aim for." (in Griffiths, 2008:103)

In this vein Louie Knight is a good person in a bad Aberystwyth. Knight grew up in the town and has a history of being victimised by those in authority – be they games teachers as a youth, or policemen as an adult ('Erw Watcyns,' said Llunos. 'He's from the Cardiff Sweeney, which means he thinks everyone else, including me, is a bumbling, carrot-crunching amateur. He's especially fond of private detectives, particularly smart ones like you'" (Pryce, 2007:130)). Knight became a private detective to try to right some of these wrongs, moved to this calling by the death of his best friend who was sent, ill, on a cross country run in the snow by the heartless Herod Jenkins:

> "For Herod the important thing was to get as many boys across the ice floes of life as fast as possible, regarding it as inevitable that some would be left behind. Who knows how differently things would have turned out if he had been a kind man? Perhaps I would be a plumber now, happy in the docile simplicity of my pipes and spanners. I would not be kept awake at night by the wolves howling in the sewer beneath the city streets, nor by the nameless anxiety that flashed through my heart every time the headlights of a turning car raked the ceiling." (Pryce, 2011:23)

Louie Knight thus becomes the moral core of the novels and the town, his integrity gives root to the humour, and gives Pryce's readership someone to root for; as illustrated in the following confrontation between Knight and a criminal in *Don't Cry for Me Aberystwyth*:

> "'But tonight it's going to be different. I'm sick of that story; I'm re-writing the script. I'm tired of risking my life on behalf of the people of Aberystwyth; trying to help them; trying to bring justice to birth; trying and failing – but failing heroically — to somehow wrest the tiller from the hands of fate once in a while and make things better

> than they should be, or than the people of this town ever have the right to expect; and doing it all for so little money I'd be ashamed if people found out. Tonight is not going to be like that. Tonight I'm going to ask you a civil question and you are going to give me a straightforward answer; the sort you would give to your mother if she asked you what you wanted for your tea. And the reason you are going to do it is simple. You will do it because it's the right thing to do; because it's the decent thing to do; because two men who have no reason to hate each other should be able to sit in a room and have a nice conversation, one in which the one helps the other for no other recompense than the joy that comes from offering a hand of help to a wayfarer lost in the gloom of this world." (Pryce, 2007:193)

Despite or perhaps because of his world-weary heroism, Louie's love life remains unrequited, and so it is left to his relationship with Calamity to give him a family and a 'home' to return to, and a place he belongs; as he narrates in *The Day Aberystwyth Stood Still*:

> "Calamity was almost eighteen now and had been my partner for five years. During that time she had become the daughter I had never had. I had felt her absence keenly. She gave up the cracks-in-the-pavement game and ran the last few steps, skipping up to me and kissing me.
>
> In acting as a father figure to her I had acquired the father's secret melancholy: watching her do her best to rush through the years of enchantment in the forlorn belief that adulthood was something worth rushing for." (Pryce, 2011:14)

Although in part a parody of 'dime thrillers' and playfully mocking the archetypes of seaside and Swansea in equal measure, Pryce's novels retain their appeal due to this genuine affection between the characters, a consistent atmosphere of humour and fate, alongside a real appreciation for the landscape of the area. I ask Pryce whether the beauty of the natural world around Aberystwyth offers a counterpoint to the murky affairs of humans and school teachers?

> "In my view, part of the reason the comic absurdity works is it because it is underpinned by a foundation of seriousness - the world is daft in many respects but the people live and love and hurt and bleed in a real way. The portrayal of the landscape is part of that, it is genuine homage to Aberystwyth rather than piss-take." (PC)

Pryce, once schooled in Aberystwyth, "can remember most places" in the area as the landscape is "deeply inscribed on my heart" (PC). A particularly evocative passage takes Louie and his friend Myfanwy (a chanteuse from the Moulin Club) in his Wolseley Hornet out of Aberystwyth on the back roads to Borth.

Figure 8.2 Looking down to Borth and Ynyslas sand dunes on the backroads from Aberystwyth. (Source: Author)

> "At the top of Penglais Hill we turned left and took the old route to Borth. The sun was hot, the windows were open and Myfanwy sang as we drove. It was like sailing a ship over an ocean of grass as the road went up and down over the hills and dales. Every hillside was chequer-boarded with cows. The constant rising and falling of the landscape had a hypnotic regularity and you thought it would never end. But then the car mounted the final hill with that suddenness that never fails to surprise and we were on the roof of the world, staring at nothing but blue: the washed-out blue of the hot sky, and the darker indigo of the cold sea rolling in from the Bay. We pulled into a driveway in front of a five-bar gate and got out. The hillside curved steeply away down to Borth and the wind was fierce, buffeting the car and making the loose cloth of my shirt flap with a sharp sound. Down below us, extending for almost ten miles, was the huge flat expanse of the Dovey estuary and stretched across it in a thin straight line was a straggle of houses. This was the town of Borth: tinselled up with inflatable swimming hoops, buckets and spades in summer; and in winter nothing but dust and creaking shutters. At the far northern extreme, lost in the haze and a desert of sand dunes was Ynyslas, goal of our picnic; and beyond that, on the other side of the estuary, were the dot-sized houses of Aberdovey. From here they looked achingly

> close, but so formidable a barrier were the estuarial tides, that Aberdovey often seemed like another country." (Pryce, 2001:59)

At the end of the back road from Aberystwyth lie the sand dunes at Ynyslas, and it is here that Louie lives in a caravan. This remote location was often a useful place for the detective to hide from his enemies, but was also simply a beautiful part of Pryce's entangled world.

Figure 8.3 Looking up the Dovey Estuary from the Ynyslas sand dunes. (Source: Author)

> "It was just after midnight, maybe 1.00 or 2.00 in the morning. I lay asleep in my caravan in Ynyslas. The far-off susurration of the waves was barely audible, but the wind coming in off the sea cuffed the caravan like the hand of a giant schoolteacher and made the metal fabric sing. There is something deeply comforting about that sensation, of feeling protected and cocooned in warmth and yet aware, too, of the proximity of the ocean. Ynyslas is 6 miles north of Aberystwyth and lies hidden from the world in a corner of sand adjacent to the estuary. During the day in summer nothing moves here except tide and cloud and, occasionally, across the estuary on the distant hill, two carriages of a toy train going north.
>
> My caravan was on the landward side of the dunes and enjoyed a view over the top of the other caravans through the netting of TV aerials to the Dovey Estuary. When you die, if you have enough clout to get in the VIP seats, this estuary is what you look at." (Pryce, 2011:55)

Box 8.2 'Not all places are as beautiful' as Ynyslas.

If you divert from the back road between Aberystwyth and Borth, you find yourself in a coastal hamlet called Clarach. As Pryce describes:

> "If Borth is the poor man's Aberystwyth - and it isn't - then Clarach is the poor man's Borth. And that's about as poor as you can get without selling a kidney. It's not a one-horse town, not even a hoof, maybe the imprint left by a horseshoe nailed once long ago to a fence or maybe just a handful of oats." (Pryce, 2005:111)

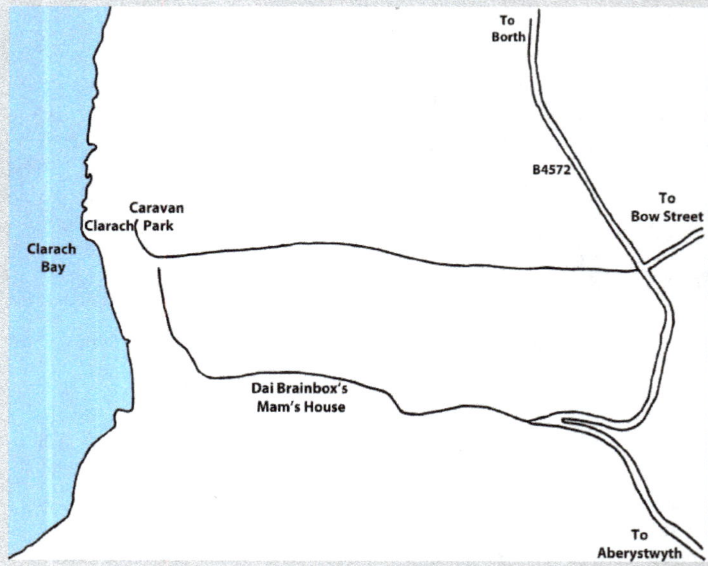

Map 8.1 The back road to Borth and the tracks to Clarach.

When I was poor in Ceredigion I lived in Clarach, and my neighbour was the villain of *Aberystwyth Mon Amour*, Dai Brainbocs. Pryce located Dai Brainboc's family house on the south side of the valley, facing north, in an area permanently shielded from the sun (see Figure 8.4). Houses there lived in sodden perma-gloom like the homeland of the Snow Queen. I asked Malcolm Pryce how he chose the inland locations for the residences of his characters, for example locating Dai Brainbocs' Mam's cottage overlooking the 'poor man's Borth'.

Figure 8.4 The valley at Clarach, looking to the frosty, north-facing slope. (Source: Author)

> "Dai Brainbocs's Mam's cottage was just something I dredged up from memory. I walked past just such a cottage once and remember thinking how it would soon crumble like cottage cheese on account of the all-pervading damp. I think I just cast around my mental map of the terrain for suitable places to set the adventures. Often suitability involves nothing more than being a short drive from the centre of town and sufficiently bleak as to be the plausible abode of a nutter." (PC)

Pryce sums up this area in the following way:

> "After a short walk over the hilltop you arrive at the top of a valley that the sun never kisses; never even shakes hands with or even acknowledges with a curt nod. On the way down you pass dung-caked sheep with contempt in their eyes. Animals so destitute they spend their whole lives on an incline, front legs higher than the back in the morning, back legs higher than the front in the afternoon, eating gorse and perhaps grass on special occasions; and they give you that look. Pity or contempt? It's hard to tell but when the sheep lets out a loud *blaaa* and a fraction of a second later the whole hillside erupts in answering bleats you can't stop yourself thinking: in your heart you know it's impossible, but it sounds suspiciously like laughter." (Pryce, 2005:111)

Figure 8.5 The grit at Clarach, with Aberystwyth semi-hidden in the distance. (Source: Author)

"This was Clarach. For entertainment you can lose some money at the amusement arcade situated in a breeze-block room that anywhere else on earth would be called a garage. Or you can take a car out to the main road and drive fast over the humpbacked bridge near the church. Or, if the melancholy fit is upon you, you can walk to the patch of grit smeared against the land's edge which the map describes as the beach and walk out to the end of the rocks and look to your left. If the tide is low enough, and you can go out far enough, you can see a little bit of Aberystwyth peeping out from behind the cliff. In that moment you regret all the bad things you ever said about her." (Pryce, 2005:112)

Llamas and lamentation: The globalisation of the Pryce diaspora

The plotlines of Pryce's novels do not simply stretch from Aberystwyth to Aberdovey, but also beyond. Pryce has created a believable 'universe' which has connections as far as Ukraine and South America (as Griffiths chuckles, in Pryce's stories "Aberystwyth's promenade is very, very long" (2008:103)). In *From Aberystwyth with Love*, for example, we see how Pryce uses Aber's diaspora to offer a fond but nevertheless sharp satire on hiraeth in the scene when Louie and Calamity are introduced to resistance fighters in Hughesovka, Ukraine. Although many may consider Hughesovka to be a fictional invention, modern day Donetsk was founded by Welshman John James Hughes in the 19th century, lending his name to the first settlement on the site (see Henry, 2013). Pryce's employment of this historical truth adds further complexity to the entanglement of reality and fiction in Louie Knight's world:

> "Calamity grinned and punched me on the arm and we both gazed out at the countryside flowing by. The gently rolling farmland of the Western Ukraine slowly gave way to the outskirts of that longed-for Eldorado, Hughesovka. ...The buildings got smaller, the tram lines ended, and we found ourselves in a wasteland of featureless modern apartment blocks.
>
> 'Wow! It's just like Penparcau[2],' said Calamity for whom all new experiences were a source of wonder." (Pryce, 2009:103)

> "'So this is the Welsh Underground.' said Calamity.
>
> There was an awkward silence after she said this. Edwards the Fascist Wrecker explained, 'We don't normally talk about our . . .our nature quite as openly, but, yes, I suppose that's as good a term as any.'
>
> 'What do you do?'
>
> 'We meet when and where we can, in safe houses mostly. We also print our own samizdat . . .' He pulled out a tattered and tightly rolled-up mimeographed pamphlet. It was called *Hiraeth* and bore on the cover an illustration of Barry Island pleasure park....
>
> 'Our aim is to return to the Motherland for which we all ache so terribly in our hearts. But, of course, the authorities will never allow such a thing. And so we try to gather what scraps of information we can. Your visit to us today is a great and longed-for honour. Occasionally we organise an escape. But the authorities are cruel and cunning and thwart us at every turn.'
>
> 'What do they do?' I asked.
>
> Lewis the Pedlar of Nationalist Opiate under the banner of Proletarian Literature replied soulfully as if he bore the pain of all the failures in his own heart. 'They intercept our escapers and arrest them. We don't see them for many months. What happens to them is not clear but we believe they are taken to a special psychiatric institution where they

[2] Penparcau – a deprived neighbourhood on the outskirts of Aberystwyth.

> are re-educated. ...After the treatment they are released back into life in Hughesovka but they are not the same as they were before. They avoid their old haunts, shun their old friends and contacts. After many months, quite by chance, one of us will run into one of them in the bread queue or something, and say, "Hi, Boris, what are you doing here, I thought you'd gone to Wales?" And he will refuse to answer, look away and mumble something sheepishly, some absurd story about how he travelled to Wales and decided to come back because he didn't think it was all that good.' The people in the room scoffed in disgust at the absurdity of the notion." (Pryce, 2009:104)

Coupled to the entanglement of Donetsk into Louie's world, Malcolm Pryce also employs the more well-known existence of Welsh enclaves in South America into his Aber-diaspora. Onto this real geographical connection Pryce fuses the 'Patagonian War', colloquially known as 'Welsh Vietnam'. We are first introduced to this entanglement through Cadwalader, a war veteran Louie meets on the golf course of Ynyslas. As Pryce recalls:

> "for some reason, I imagined a veteran of the fictitious Patagonian War searching for golf balls in the rough. From that one thought it all flowed. But I probably wouldn't have imagined the war veteran if it hadn't been for the bleak, beautiful and strangely forlorn character of Ynyslas and the landscape round there." (PC)

> "We caught the train back to Borth and took the bus outside the station, down the arrow-straight road that bisects the golf course, towards Ynyslas. On Calamity's lap was a bag from Peacocks containing the sort of coat that used to be popular with medieval Jews and that had now come back into fashion with druid assassins. Meirion said he thought it had something to do with the military. And in West Wales there is only one military. The Welsh Foreign Legion, famous or notorious depending on the way you looked at it, for the campaign to liberate the former Welsh colony of Patagonia in 1961." (Pryce, 2003:93)

> "It was all more than a quarter of a century ago now, but for the army of broken ghosts that haunted the fields and lanes of West Wales the memory burned as fiercely as ever. Five minutes in a recruiting office above Boots was all it took to seal their fate – farm boys who'd never been further than Builth Wells rubbing shoulders with a rag-bag of foreign intellectuals, artists and soldiers of fortune. Five minutes to think of a *nom de guerre* that hadn't already been taken and sign your cross on the dotted line and head off to the Welsh Vietnam." (Ibid.:93)

> "Fifty yards ahead of us a lone figure could be seen tramping through the knee-high grass. His tattered army greatcoat and forlorn demeanour marked him out as one of the veterans from the war.... Patagonia, a harsh tract of land on the tip of South America, a world of searing beauty and withering cold; difficult to find on an atlas and known only because Welsh settlers went out there in the nineteenth century. A story that began in adversity and ended in tragedy seventy years later when the Indians turned against them. It was a war of independence that soured a generation and left behind the legacy of the Vets: [e]ach carrying in his heart the story of a military adventure that no one wanted to hear.
>
> He was looking for lost golf balls which he could sell for his evening's meal. There was a sudden shout, a sharp crack, and the old soldier spun to the ground; felled by a golf ball. We ran towards him and from the fairway the party responsible for his misfortune came over at a more leisurely pace." (Pryce, 2001:60)

Box 8.3. Beyond the bars of vanilla prisons

The trajectories which shoot off from Pryce's pier also now include minor characters such as Sospan the Ice Cream seller. Whilst between the pages of Pryce's plotlines, Sospan is marooned "behind the invisible bars of his vanilla prison" (Pryce, 2001:217) beyond those pages (through Pryce's new website), Sospan has a new lease of life where he can experiment with ice cream flavours advertised in our own 'reality'.

> "Did it perhaps resemble a starfish, or a fairy toadstool? A goblin's hat, a vampire's tooth or a Schoolgirl's nipple?'
>
> 'Schoolgirl's nipple,' I said.
>
> Sospan smiled as if that were his favourite and then said, 'I've just had a brilliant idea.'"
>
> Sospan has released his new season's flavours, a bit late this year. They are more than usually daring and avant-garde. I think he is a bit miffed by the fuss the newspapers made last year about Heston Blumenthal serving up egg and bacon ice cream, even though Sospan has been offering it for years. ...Anyway, here are the flavours:

Ambergis, Ectoplasm, Sebum, Ecstasy, Agar jelly, Millipede, Eucharist, Placebo, Strawberry, Embalming Fluid, Soot." (see Pryce, 2013)

Conclusion

I ask Malcolm Pryce where's the best place in Swansea to get an Italian suit with double vents?

"I have been told in no uncertain terms by the mayor of Swansea to keep away or I will end up sleeping with the fishes." (Pryce, PC)

Like an detective's old trenchcoat, Louie Knight's Aberystwyth has taken on a personality of its own. Aberystwyth has taken on a (double) life; this remote, sleepy, sleeper town has been woken up by Pryce and turned into a film noir, with its own back- and cover- story. This poetic refraction is based on the quirky reality that constitutes Aberystwyth, that at the end of this line what is real, what is imagined, and what is oblique parody becomes increasingly difficult to discern. Riven with self-mockery but underpinned with affection, through Pryce's plotlines Aber becomes a place where its inhabitants cry, love, and muse on the incredible lightness of being, in the interstices between past, present, genre and geography.

Echoing themes in Chapter 7, Malcolm Pryce's (youthful) relationship to the place of Aberystwyth stayed in his heart and mind and has become material for his active imagination. Through so doing, it has resonated with others' recollections of the town – as well as introducing the unfamiliar to the area – and entangling them into his Aber-diaspora. Through his Louie Knight series Pryce has added new stories/storeys to Aberystwyth. His books function as portals through which the reader can pass through the 'real' and 'fictional' versions of the town. Adding an affectionate and thrilling element to the place, the effectiveness of his stories/storeys mean that readers choose to blur the distinctions between them, so like a stick of rock candy, his words permeate and come to define the context they describe. Pryce has created a world that it is possible to enter via the turn of a corner or the turn of a page, and the real becomes (un)hinged by the escapades of mafia druids, evil schoolmasters, and moulin girls in stovepipe hats.

Pryce's Aberystwyth offers an example not only of how it is impossible to disentangle the (con)fusion between the page and place, but also how a sleepy Welsh seaside town has inevitable connections across the globe, with past inhabitants, war veterans and readers tying the apparently insular, past, and remote, into the international present. It is in this entanglement between page and place that Louie and Calamity seek to bring justice to the old courtesan that is Pryce's Aberystwyth:

> "I hugged her and told her how glad I was that, despite the truly terrifying odds thrown up by the universe, Calamity and I happened to be sharing the same planet, the same epoch and, best of all, the same office. And in a region of the solar system where the rain seldom lasts for more than a week. She pressed her face against me and spoke into the folds of my trenchcoat... I laughed, and together we walked slowly off into the wind that never stops blowing." (Pryce, 2011:244)

Part III

Nations

Chapter 9
Reading gave me worlds: Gillian Clarke's autobiographical plotlines

If Peter Finch is the Bard of Cardiff (Chapter 1), Gillian Clarke is the National Poet of Wales. Clarke's plotlines, like the conversations of her neighbourly "hedge poet" (Clarke, 2008:6), run from the local and the familiar, over hills sides and regions, and contribute literary lanes that criss-cross the Welsh nation. Often considered as a feminist writer (see Peach, 1993), this chapter suggests that Clarke's plotlanes/lines can also be usefully understood by drawing on broader traditions of nature writing. As Clarke herself states, "[t]here are often verbal treasures to be found in the language of geology [and] geography, a fresh vocabulary and a startling imagery" (Clarke, 2008:41). I argue here that this sensitivity to the language and consequences of geography repositions Clarke as the Welsh incarnation of North American writers such as Barry Lopez (1993; 1998; 2001; 2004), Gary Snyder (1974; 1983; 1990; 1995; 1999), and Annie Dillard (1975, 1982, 1999).[1] Although, as Finch acknowledges, the "wide open space" of the United States cannot be directly transposed to the lanes that go "up valleys and [get] lost among pits and sheep and trees" (Finch et al., 2006); Clarke nevertheless captures the "solace" of Gretel Ehrlich's "open spaces" (1985) and evokes the umbilical connection between Welsh people and their places. In this chapter, instead of following a bespoke plotline through the work of Gillian Clarke,[2] I draw together the entanglements between Clarke and this

---

1 When I ask Gillian Clarke about the resonances between her work and these writers, she answers with respect to Annie Dillard in particular, "I love her work. I met her before I read her, in the 1980's, and we walked a mountain in New England and picked up a garter snake. We got on very well" (PC).

2 Clarke offered an invitation to her home to walk and talk in her local countryside, however the life of a national poet meant that deadlines, festivals and publishing got in the way. This, coupled to a dislike for interviews – as she states: "I am ALWAYS disappointed with transcribed interviews. I am usually mis-quoted, although it is probably my fault for talking too fast!" (PC) – meant that correspondence was limited to email.

tradition in order to shed new light on the important aspects of belonging and connection that she engages with in her work.

Big and little: Exploring the "complex experience of real life tangled with the life of the imagination" (Clarke, 2008:32)

As we will see, Gillian Clarke's body of work engages with key debates that question the future of the Welsh nation. However, she achieves this through focusing not on the macro- or meta-scale of inquiry, but through focusing on the small and immediate. By zeroing in on the here and now, on the hearth and home, Clarke is able to collapse the world to a scale that is meaningful and accessible. As she states:

> "This is … the sort of vision we need right now[:] slow down, focus, consider one small detail in the great complexity that is the world. [This will] remind… us why we live here." (Clarke, 2008:135)

Including writers such as William Blake (1863), the ability to see the 'world in a grain of sand' is an attribute associated with many great poets. Iain Sinclair (see Chapter 12), when travelling to meet Gary Snyder on the west coast of the United States, remarks on a similar trait in the American nature-writing tradition:

> "A hundred acres of ground is a substantial mind-map for a poet. Snyder's friend and colleague Lew Welch saw eternity in an inch of lichen on a rock. 'These are the stamps on the final envelope,' he wrote." (Sinclair, 2012a: no page)

This awareness of the timely and eternal existing in the highly localised here and now is a poetic sensibility Gillian Clarke developed from a young age, as she outlines in *Source*:

> "As a child I used to play a game which I called 'big and little', which now seems to me a primitive version of a poet's game, physical and imaginative in nature, yet a child's way into a questioning habit of thought. Half-close your eyes and stare, or blur your ears. A stone becomes a planet. Your breath is the wind, a quarrel is a storm, a storm becomes a war. It works the other way too. Your cupped hand can balance the pebble of the setting sun before it is dropped into the sea. With a finger you can blot out a Neolithic stone, or a planet. Take a magnifying glass to your thumb-print. Place a hair under a

> microscope. These are geographies. It is a game played with scale and perspective that has always fascinated me." (Clarke, 2008:31)

The 'primitive... poet's game' of 'big and little' draws attention to the arguments of philosophers such as James Gleick who suggest that all relations are 'fractal' in nature, or in other words, that large scale relations are magnifications of the same delicate tensions between order and disorder which exist at much smaller scales (1990). As Gleick states, for example, "in reality, a river's basic shape... is not a line but a tree. A river is, in its essence, a thing that branches. So are most plants.... So is lightning. So is the human lung. ...In the vernacular of science, it is fractal, its structure echoing itself on all scales" (1990:no page). Clarke's game of 'big and little' not only highlights the fractal nature of existence but also reminds us of what can be gained from witnessing the physical processes of nature. In this latter insight we can see the importance of the relations between people and place, of how everyday experiences even at a young age connect individuals to the world around them, and come to define in small (or little) but important (or big) ways their identity and outlook. As if inspired by the simple logic of this children's game, Gillian Clarke's work explores how our identities, both personally and collectively, are built up in delicate tension with places, and how places in turn contribute to the generation of our world view and soul. Clarke, like Lopez, Dillard and Snyder, warns us of the risks we take if we become oblivious to these person-place relations, and her work in particular raises questions for Welsh culture about what it may lose if it loosens itself from its historic roots. Clarke's writing therefore serves as an important reminder of how culture is rooted to the land and how only through understanding these relations can we make informed decisions about the nation's direction to the future.

The roots of nation building: auto-biographical plotlines

The geographical sensitivity that led Clarke to consider the abiding relations between people and place was informed not only by the childhood games of 'big and little', but also by the entanglement of page and place that formed the auto-biographical plotlines of her youth. Through combining overheard conversations with family stories, the fusing of the real and the imagined became definitive of her worldview:

> "I eavesdropped on grown-up talk in Welsh and English, voices over my head which seethed with rumour, gossip, quarrels, scandal, malice, stories, the Bible, Grimm, Hans Andersen, the *Kathleen Fiddler Omnibus*. My father's stories on the slow road between Cardiff and Pembrokeshire as the danger of bombs came and went. Words. Delicious words. Names. Coed ar hyd y Glyn. Splott. Golate. Brains Dark. Coedpoeth up north where Nain and Taid lived, and Gobowen, and Wrexham Lager Beer. Once at Fforest, balancing on the slippery wall that dammed the flow of slurry between the shippon and the sloping yard, I was muttering favourite words, collected into a couplet, stamping its syllables as I went:
>
> Ga puts mentholatum on her sciatica
>
> And Ceri soaks the clothes in Parazone." (Clarke, 2003:56)

Through childhood adventures, family trips, world events, and narrated stories, the entanglement of real and imagined, page and place, were absolute. For Clarke, real life was "mediated, modified and deepened by literature" (Clarke, 2008:77), and the distinction between fact and fiction, there or here, was collapsed, as Clarke articulates with reference to a tragic news story of the time:

> "Two … children were found half-buried in leaves in a wood. *The Mirror* or *The Graphic* or *The News of the World* headlined the story 'Babes in the Wood'. My mother hid the paper but I saw it. Story became life, life became story. It happened far away, but for me it was Porthkerry Woods, with bluebells and the sound of the sea, a train crossing the viaduct and my feet running down dry woodland paths and out over pebbles in broad sunlight where my mother was unpacking the picnic." (Clarke, 2008:59)

In this way imagination, geography, fiction and fact lay together in Clarke's mind to become part of her being. Through reading, remembering and misremembering literature, Clarke created worlds that came to inform her perception and outlook on nature and nation:

> "The fact that literature, from nursery rhyme and fairy story onwards, was so closely associated with the natural world, has played a strong part in making me a country person in my head rather than an urban one, even during the years of my life growing up in the city, and bringing up my children there. Literature hallowed the natural with the supernatural. It made the stones sing. It populated the countryside with animals, seen and unseen. It made natural phenomena reverberate with

mythological meaning, turned a rocking stone to a giant's apple, a rock pool to a footprint." (Clarke, 2008:11)

'Dyma cariad fel y moroedd', 'Here is love like the sea', or the hauntingly simple poetry of 'There is a green hill far away', came the equally strange and beguiling spells of Shakespeare, spells such as 'Nothing will come of nothing. Speak again'. I have not, nor will I check these quotations from the Bible, the hymn book, and from Shakespeare. This is how I remember them. This is how the words must stay, even, perhaps, in misquotation. Those words, in that order, are a part of the 'fiction' I have been writing all my life." (Clarke, 2008:32)

Box 9.1 Episodes in plotlines

In *Source* (2008), Clarke articulates her route to adulthood and the role that geography played in locating and informing her sense of belonging. As we have seen in Chapter 1, in her childhood Clarke's father told her stories that became ties into local places,[3] and as a student new person-place relations were developed through literature, practice, and politics in the capital:

> "It was obligatory to wear academic gowns to lectures and in town, so we were a conspicuous group. Cardiff belonged to us. The places where we consorted, flirted, romanced, sorted the world out, planned the anti-Suez march, anti-war or anti-nuclear demonstrations – the Students' Union, the Kardomah, the Hayes Island Snack bar, the pubs, the Old Arcade, the 'Woody' – put a new, grown-up Cardiff under my skin and into my bloodstream." (Clarke, 2003:62)

From that chapter in her life, Clarke returned to live in a flat in the family house. Echoing themes from Hadley (Chapter 5) as well as Williams (Chapter 3), she experienced the power of place in influencing memory and identity:

---

3 As we saw in Chapter 1, Clarke notes how "My father set the stories of the *Mabinogi* in locations wherever we happened to be - in Fforest, in Cyncoed, in Kingsway, in Roath Park. ... these were the first stories located in a geography that was mine" (Clarke, 2003:56)

> "Our house in Cyncoed was 'the first house on the meadow'. An elderly neighbour told me so when, years later, at twenty-two newly married, I returned there to live in the upstairs flat. No other house has established itself in such detail in my mind. I lived there at crucial periods of my life, times when heightened awareness of the indoorscape lays down rich memories: earliest childhood; the first years of marriage; the birth and upbringing of three children. ... 1 Cyncoed Avenue was familiar in the complete meaning of that word, a house I can never leave behind. I recorded this in a sequence of poems called 'Cofiant' ('Biography'):
>
> Houses we've lived in inhabit us
>
> and history's restless in the rooms of the mind.
>
> It works both ways. I can enter that house right now, because it is a part of me:
>
> How can you leave a house?
>
> Do they know, who live there,
>
> how I tread the loose tile in the hall,
>
> feel for the light the wrong side of the door,
>
> add ray prints to their prints to my old prints
>
> on the finger-plate?" (Clarke, 2003:54)

In this passing through and passing on of places to others, Clarke gains a perspective on the importance of legacy to place – on what we must preserve and what we must allow to be re-created. She is aware, for example, that Cardiff is "my daughter's city now, and her children's" (Clarke, 2003:63), and this generational perspective, along with her geographical sensitivity, allows the exploration of larger questions of belonging in the twenty-first century. To borrow Charlotte Williams' words, it is as if Clarke is contemplating both a personal and a collective question of: "what exactly must [I] pass on…" to future generations, cautiously considering what "the essence of my generation [is] that may be meaningful" to the next (2002:190). Through her work, Clarke seeks to "piece together the parts of the collective memory" (ibid.) that may hold the nation together in the future.

"But I've no spade to follow men like them. / Between my finger and my thumb / The squat pen rests. I'll dig with it" (Heaney, 1966:5)

In order to appropriately plot a course from the present to the future, it can be argued that Clarke considers it vital to know where we have been and where we want to go. For Clarke ancestry and history, both co-constituted by the land, is a crucial co-ordinate in this trajectory. Uncovering ancestral connections has always been important for Clarke in order to understand her positioning both as an individual and as part of a nation:

> "When you write… you release something that's actually pretty deep within your memory and imagination. …I realized that digging deep was an important way forward for me. Going into the past, going deeper, going down through the layers became important. I was mining my own memory, my family's memory and Welsh memory as far as I could… That is how I am writing now, because the past informs what we are now." (cited in Peach, 1993:79)

Through writing Clarke therefore performs what has been a vital act in Welsh culture and history: mining. Although like Heaney she has "no spade to follow" the coal mining tradition, through her writing she nevertheless performs metaphorical digging into the past. Through this process she cuts into the geology and history of the country, confirming as she does so the primal connection between people and the landscape. This is particularly evidenced in the poem 'The City', which outlines a journey through Wales into England and the changing histories and cultures that are associated with this trajectory. In this edited extract we see the personal and collective histories that are written into the land, in particular the border crossings of people in order to generate a livelihood, serve industry and empire, and how roots can be misplaced through this process. This journey ends in the present at the Millennium Centre, Cardiff Bay (at the 'music palace built on bedrock' (see also Finch, Chapter 2)), and at a (geological) crossroads. It is as if Clarke is asking, what is Wales to become in the twenty-first century?

> The City
>
> The Journey
>
> A four-hour drive through geological time, to the two blue notes of the cuckoo: C and A flat again and again from oak woods sweet with bluebells as if it sang the scent and colour of blue.
>
> And in my mind a sedimentary language, words laid down on the mind's bedrock, names remembered from an old geography book: Ordovician flags, shales, mudstones, Silurian shale, grit, greywacke, limestone formed from crushed fossils, dead reefs of coral, and the petrified stems of sea lilies from warm and long retreated shallow seas; Cambrian slate that once roofed the nation; the coal-bearing shales and limestones of the Dee, where my grandfather farmed stony land and milled his corn to the west of Offa's Dyke.
>
> Over the Manchester Ship Canal, over Irwell, Medlock, Tame and Irk, braided streams of Celtic and Old English, into the ancient Kingdom of Elmet where Celts fell to Angles in the last lost battle.
>
> …
>
> For Empire. For profit. For cotton. And the century of toxins and disease was in full spate. The map of the world stained with the colour of possession, the air venomous. The planet will begin to die.
>
> …
>
> My mother in the '30s, a country girl, bright daughter of a North Wales tenant farmer, learning the midwife's art in the slums of Salford, delivered babies on filthy mattresses, with nothing but old newspapers for sheets and not a shovel of coal for the grate.
>
> ...The anthems and flags of difference to be shared, the red and the green, the dragon and Union Jack.
>
> They quarrelled, got drunk, prayed to many gods, grew restive in their chains and lost their histories in the century's confusion. Till here we are at the end of the millennium, beginning again on a geological fault, holding our breath for peace, to see a music palace built on bedrock." (Clarke, 1998:52/53)

In this reading, such personal trajectories through geology and geography prompt prominent questions about nationhood and nationality. What are the histories we should reclaim from confusion, what affiliations and responsibilities should we accept, what can we '[n]ever leave behind', and what should we leave in a 'skip in the road' (Clarke, 2003:63)? For Clarke, acknowledging geographical connection is the first vital step in order to create a coherent sense of collective identity, as Peach confirms:

> "A sense of place is, of course, crucial to [Clarke's] sense of identity. As Nicholas Jacobs has pointed out: 'a sense of identity is hard to envisage without the co-ordinates of existence afforded by the consciousness of a place in space and time'." (1993:85)

Peach continues to cite Jacobs when he states: "it is through the interpenetration of tradition (time) and locality (space) that a community experiences its identity" (ibid.), and importantly, "it is when a loss of shared background becomes apparent that the need arises to formulate and to cherish a conscious idea of the tradition" (Peach, 1993:85). As 'The City' demonstrated, through the effects of mobility and empire, the experience of a shared background may become lost or confused in the twenty-first century, and through her poetry, Clarke can be seen to be 'formulating and… cherishing' a new tradition. These issues of nation building, at once both big and little, are those that Dedalus suggests puts "daunting pressures and responsibilities on anyone who would risk the name of poet" (cited in Heaney, 1980:60). However such attempts at formulating tradition are part and parcel of the nation building exercise. Nations are not natural entities, but are created through acts of politics and culture. As James Anderson puts it:

> "We tend to see national identity as natural almost like the air we breathe, when in fact is a very recent phenomenon in human history. For most of history people identified with much smaller cultural communities than modern states... they identified with their local village, their parish or their city." (James Anderson, 1992)

Another 'national' scholar Benedict Anderson argues that due to the vast scale of many countries, nation states need to be 'imagined' into being in the minds of their citizens (1991). As Jacobs identifies, formulating and cherishing traditions are fundamental to this imagining process and we can see Clarke's work as part of this move. Through her 'squat pen' Clarke is attempting to imagine a particular sort of nation into being, one with a geographical sensitivity and an appreciation for history and place. In so doing her aims resonate strongly with the geographical writings of Barry Lopez, as the following excerpts, first from Lopez then from Clarke, illustrate:

> "Geography, the formal way in which we grapple with this areal mystery, is finally knowledge that calls up something in the land we recognise and respond to. It gives us a sense of place and a sense of community. Both are indispensable to a state of well-being, an individual's and a country's." (Lopez, 1998:143)

> "And as imagination bodies forth
> The forms of things unknown, the poet's pen
> Turns them to shapes, and gives to airy nothing
> A local habitation and a name." (Clarke, 2008:37, and Shakespeare's A Midsummer Night's Dream, Act 5 Scene 1, lines 15-20)

Following Shakespeare, creating a culture connected to the land through poetry 'gives an airy nothing / a local habitation and a name'. Such habitation, or 're-inhabitation' to use the words of Gary Snyder, seeks to create a sense of belonging through taking responsibility and care for a place. As Snyder puts it,

> "re-inhabitory refers to… persons who …turn back to the land, back to place. This comes for some with the rational and scientific realization of interconnectedness and planetary limits. But the actual demands of a life committed to a place, and living somewhat by the sunshine green-plant energy that is concentrating in that spot, are so physically and intellectually intense that it is a moral and spiritual choice as well." (2000:190)

For kindred spirits like Barry Lopez, such habitation is vital in order to form the 'bedrock' for any battle to preserve land and foster local and national identity, as he states:

> "If a society forgets or no longer cares where it lives, then anyone with the political power and the will to do so can manipulate the landscape to conform to certain social ideals or nostalgic visions. People may hardly notice that anything has happened, or assume that whatever happens … is for the common good. The more superficial a society's knowledge of the real dimensions of the land it occupies becomes, the more vulnerable the land is to exploitation, to manipulation for short-term gain." (Lopez, 1998:138)

Clarke's attempt to encourage a collective re-inhabitation of the land at a national scale is not only practiced through mining geology and geography (as illustrated in the 'The City'), but also through creating personal connections to the land. After leaving the city in which she grew up (see Box 9.1), Clarke grounded her adult life in Blaen Cwrt, Ceredigion. She describes this place in the following poem:

"You ask how it is. I will tell you. ...
The wattle and daub
Chimney hood has decayed away, slowly
Creeping to dust, chalking the slate
Floor with stories. It has all the first
Necessities for a high standard
Of civilised living: silence inside
A circle of sound, water and fire,
Light on uncountable miles of mountain
From a big, unpredictable sky,
Two rooms, waking and sleeping,
Two languages, two centuries of past
To ponder on, and the basic need
To work hard in order to survive." (Clarke, 1978:15)

In Blaen Cwrt the principles of re-inhabitation are evident: a sensitivity to place (including both the manufactured and the mountainous), the accreted histories that such social and geological places leave in a location, and an elemental need to engage with these in order to build a life (see also Williams, see Chapter 3). Through her own re-inhabitation, Blaen Cwrt becomes part of Clarke's 'lifescape'. As Convery et al., explain, the concept of lifescape captures the often:

> "taken for granted social, cultural and economic interactions between humans, livestock and landscapes' ... [Through the lifescape] the environment enters actively into the constitution of persons; there is a mutually constitutive interrelationship between persons and environment." (2005: 100-101)

Since moving into Blaen Cwrt, Clarke's life has indeed become bound up with the changes and fates of the local area, as she writes in *Source*:

> "Until 1992 the land was not ours, and this room not built. At first we dreamed of owning just one small field. Cae Blaen Cwrt. It lay beyond a wired up gap in the hedge bank, a sunlit three- quarters of an acre glimpsed like the secret garden through the door in the wall. ...The land on this side of the valley was categorised as 'less favoured land'

> and our neighbour was happy to sell us eighteen acres. Much of it is steep, wooded, beautiful: a few fields, gorse-clad slopes, and the most beautiful bluebell wood we know. On New Year's Day 1993 we walked our land for the first time. In February 1994 we walked the fields to check our first eight Beulah Specklefaced ewes, took them hay, and waited nervously for their lambs to be born." (Clarke, 2008:79)

As a consequence of her reinhabitation of the land and the creation of a personal lifescape, this place (re)generated in Clarke a connection to the rhythms and processes occurring around her, in Heaney's words,

> "the features of the landscape [became] a mode of communion with a something other than themselves, a something to which we ourselves still feel we might belong." (1977:86)

This process of lifescape generation is also reflected in the writings of Ehrlich (1985) and Dillard (1975) in North America, as we can identify in the following excerpts:

> "The seasons [have become] a Jacob's ladder climbed by migrating elk and deer. Our ranch is one of their resting places. If I was leery about being an owner, a possessor of land, now I have to understand the ways in which the place possesses me. Mowing hayfields feels like mowing myself. I wake up mornings, expecting to find my hair shorn. The pastures bend into me; the water I ushered over hard ground becomes one drink of grass. Later in the year, feeding the bales of hay we've put up is a regurgitative act: thrown down from a high stack on chill days they break open in front of the horses like loaves of hot bread." (Ehrlich, 1985:90)

> "I come to this island every month of the year. I walk around it, stopping and staring, or I straddle the sycamore log over the creek, curling my legs out of the water in winter, trying to read. Today I sit on dry grass at the end of the island by the slower side of the creek. I'm drawn to this spot. I come to it as to an oracle; I return to it as a man years later will seek out the battlefield where he lost a leg or an arm." (Dillard, 1975:18)

Blaen Cwrt thus became Clarke's place of re-inhabitation, her place to create meaningful connections between place, history and identity. Through this process Clarke created her own lifescape, a plot entangling the personal to the geographical, the human to the non-human, and the philosophical to the natural. For Barry Lopez, such as plot is understood through the influence of the Spanish tradition in the term 'la querencia' (1992). La querencia goes further than the notion of the lifescape (and comes closer to Buttimer's concept of the 'lifeworld'(1976)) by confirming that our relation to the world around us includes:

> "a place on the ground where one feels secure, a place from which one's strength of character is drawn. …The idea is quite beautiful – a place in which we know exactly who we are. The place from which we speak our deepest beliefs." (1992:39)

For Lopez and Clarke, from this 'little' lifescape it is possible to realise 'bigger' lessons that may inform a consciousness beyond the individual or locale. This is evidenced in the following passage from another North American nature writer, Edward Abbey (see also Griffiths, Chapter 6):

> "A weird, lovely, object out of nature like Delicate Arch has the curious ability to remind us – like rock and sunlight and wind and wilderness – that out there is a different world, older and greater and deeper by far than ours, a world which surrounds and sustains the little world of men as sea and sky surround and sustain a ship. The shock of the real. For a little while we are again able to see, as the child does, a world of marvels. For a few moments we discover that nothing can be taken for granted, for if this ring of stone is marvellous than all which shaped it is marvellous, and our journey here on earth, able to see and touch and hear in the midst of tangible and mysterious things-in-themselves, is the most strange and daring of all adventures." (1975:37)

For Lopez, and by extension also for Clarke, the strength of the notion of 'querencia' is the strength it affords to its communities when they are under threat. As Lopez goes on to explain, the idea of querencia is thus vitally important in the contemporary context as it:

> "carries this sense of being challenged … by something lethal …I would like to ... suggest it [also] applies to our challenge in the modern world, that our search for a querencia is both a response to threat and a desire to find out who we are. And the discovery of a querencia, I believe, hinges on the perfection of a sense of place. A sense of place must include, at the very least, knowledge of what is inviolate about a relationship between a people and the place they occupy, and …how the destruction of this relationship, or the failure to attend to it, wounds people." (1992:40/48)

Clarke's Blaen Cwrt is at once her *Source* – her querencia – where she seeks to determine her own identity and through this, looking larger to formulate a new geographical 'nation'scape for Wales itself.

Box 9.2 Clarke's Source and nature writer entanglements

Clarke writes of her experiences of re-inhabitation and lifescape in the book (titled perhaps to echo the sentiments of la querencia): *Source*. In many ways it is written as a journal, in the same tenor and tone as Aldo Leopold's *Sand County Almanac* (1949) or *Round River* Journals (1953). In these, both Clarke and Leopold mark time in their lifescape, as Clarke writes:

> "On the first day of January I open a new journal and mark the clean page with a date, location, a first sentence. New Year's Day 2007. It is always an unlined, hardback black book, three inches by five, with acid-free paper. The first words print the field of snow. I relish the moment, the pen, the white page." (2008:67)

> "The weather-people keep talking of a once-in-twenty-years hard winter, like 1947, '62, '83, coming any year now. Such talk makes my heart beat faster. The drama of extreme weather is both fearful and thrilling. These past few years it has come to nothing more than a few weeks of dry, icy weather. Usually frost is kept at bay in the west by the temperate air of the Atlantic, but when winter anti-cyclones stall over the North Sea, the Atlantic holds its breath. Snow falls to the east, while we lie under iron frost, snowfall held back by the hills." (2008:67/68)

In this way Clarke marks Erhlich's 'Jacobs ladder' of the seasons, and in her journal the thoughts of the year become accumulated, she (like Dillard) "read[s] and write[s], and things I have never understood become clear; I reap the harvest of the year's planting" (Dillard, 1975:45). *Source* therefore accumulates the insights gained from her engagement with Blaen Cwrt, from the "recognition of ethical and aesthetic values to be found in natural things" (Leopold, 1953:VIII). As a consequence it is possible for the reader to sense the atmospheres of when and where the diary was composed, the entries have "the pungency of oak smoke", and convey "some measure of insight into the... field experience from which perception gradually emerged" (Ibid.).

> "I am eye to eye with the blackbird. She bustles, foraging on the ground. A kite flies low and slow over the field. Between the branches of the beech trees is a glimpse of our eight pregnant ewes, all descendants of our first little flock, grazing the top corner of the Fron. Sunlit reflections from the 'moat' run up the oak mullions between the glass. In rain the water sings and syncopates in the down-pipes. After dark the glass walls are glossy against the black of the night, and the glow from this room signals across the fields." (2008:80)

Through *Source* we see how it is possible to become sensitive to and part of a lifescape which deepens and enriches our sense of co-ingredience with the natural processes that are part of our world. In this way, Clarke's writing resonates strongly with North American nature writers like Thoreau and Dillard who have also undergone a process of re-inhabiting the land:

> "Every day a new picture is painted and framed, held up for half an hour, in such lights as the Great Artist chooses, and then withdrawn, and the curtain falls.
>
> And then the sun goes down, and long the afterglow gives light.
>
> And then the damask curtain glows along the western window.
>
> And now the first star is lit, and I go home." (Thoreau, cited in Porter, 1962:156)

> **"Heron at Port Talbot**
>
> Snow falls on the cooling towers delicately settling on cranes. Machinery's old bones whiten; death settles with its rusts, its erosions.
>
> Warning of winds off the sea the motorway dips to the dock's edge. My hands tighten on the wheel against the white steel of the wind.
>
> Then we almost touch, both braking flight, bank on the air and feel that shocking intimacy of near-collision, animal tracks that cross in snow.
>
> I see his living eye, his change of mind, feel pressure as we bank, the force of his beauty. We might have died in some terrible conjunction." (Clarke, 1982:50)

> "But I was standing on the bridge, not sitting, and it [the muskrat] saw me. It changed its course, veered towards the bank, and disappeared behind an indentation in the rushy shoreline. I felt a rush of such pure energy I thought I would not need to breathe for days." (Dillard, 1975:171)

A Welsh Querencia

Although it has been argued that Clarke can be seen to be entangled in spirit and substance with the nature writing tradition that has been cultivated in North America, Clarke is also doing something particularly Welsh. Through her body of work Clarke is putting words to the notion of hiraeth in order that it can be recognised and valued in the twenty-first century. If Wales is at risk of losing its identity in the maelstrom of globalisation, then Clarke appears to be locating a querencia in order to remind the nation what it may be giving up if it does so. Through her writing Clarke is, in effect, asking the question: how we should live? If, as Dillard (1982) suggests, death will come to us all, the only question that matters is how best to live, as both individuals and as a nation:

> "We can live any way we want. People take vows of poverty, chastity, and obedience – even of silence – by choice. ...I think it would be well, and proper, and obedient, and pure, to grasp your one necessity and not let it go, to dangle from it limp wherever it takes you. Then even death, where you're going no matter how you live, cannot you part. Seize it and let it seize you up aloft even, till your eyes burn out and drop; let your musky flesh fall off in shreds, and let your very bones unhinge and scatter, loosened over fields, over fields and woods, lightly, thoughtless, from any height at all, from as high as eagles." (Dillard, 1982:16)

In this reading, for Clarke her querencia is her one 'necessity', and perhaps for the nation the connection to landscape is equally essential. As evoked in '*Snow on the Mountain*' (1978), Clarke implies that the nation is tracked by a 'black crow', the identity of the country of her birth may (like the trees in the '*Cardiff Elms*' (1982)) be facing a 'terrible undoing' (or like the passengers in '*Railway Tracks*' (1978), be 'losing its way'). This undoing may be in terms of it culture, its practices, or its voice. The language that found 'its tongue in the ancient dark of the deepest aquifers' (*Language Act Eisteddfod,* Clarke, 1998) and whose 'cadences shaped / By the long coast of the peninsula [and] continuous pentameter of the sea' (*Llyr*, Clarke, 1982), may 'Crumble... to wind and bird-call' (*Clywedog,* Clarke, 1978*)* by the threats of modern life. In a way, like the 'ancient priest' of 'St Augustine's Penarth', Clarke's words are 'preaching continuity / in the face of turning tides' (1978).

Re-inhabitation is thus a retrenchment to defend the nation from threat, such as that depicted in *Jac Codi Baw*(Clarke, 1992). As Peach argues, "what emerges from *Jac Codi Baw*...is an image of the statue-of-identity holding the scales of progress in which the 'past' and the 'present', 'demolition' and 'preservation', are in equal balance. *Jac Codi Baw*— a local Welsh-language nickname for a J.C.B. ... contrasts the time it has taken to tear down the building with the time the building has stood and the time it has taken her, as a shopper, to fill a shopping bag" (1993:80). As Clarke writes:

> ".... We are used to the slow
>
> change that weather brings, the gradual
>
> death of a generation, old bricks
>
> crumbling.
>
> [but]
>
> They have torn down in the space of time
>
> it takes to fill a shopping bag,
>
> the building that stood beside my car.
>
> It was grown over with ragwort,
>
> toad flax and buddleia, windows
>
> blind with boarding...." (Clarke, 1992:38)

Again resonating strongly with Dillard, Clarke is calling us to pay heed to such changes, as "it is difficult to undo our own damage, and to recall to our presence to that which we have asked to leave. It is hard to desecrate a grove and change your mind" (1982:70). If such irrevocable changes are made in haste, they are liable to cause damage and leave scars; as Clarke writes in *The Stone*, once changes occur, culture, like the sea, may "still fizz over a bruise that will not heal" (see Clarke, 2008:38). In *Storm Awst* it is possible to frame the small holding (or tyddyn) as a lifescape metaphor for the country as a whole as it faces the August Storm[4] of change from English language and international culture.

> "... A curious
>
> Calm, coming from the storm, unites
>
> Us, as we wonder if the work
>
> We have done will stand.
>
> Will the tyddyn
>
> In its group of strong trees on the high
>
> Hill, hold against the storm
>
> Awst
>
> Running across hills where everything
>
> Alive listens, pacing its house, heart still?" (Clarke, 1978)

The threat from England (directly described in *Fires on Llŷn*, 1998), and the broader internationalised culture that England also represents (see Ritzer, 2002) is hinted at geographically in *Return to Penarth* (Clarke, 1978). Here Clarke draws attention to the coastal homes surveilling the horizon for enemies abroad: "These houses / Keep their watch, old and half-blind / Wall-eyed and white with premonitions / Of rain from England" (Clarke, 1978). The constant presence and threat of international culture brought by the tide and weather can be seen to erode history and culture from the land;[5] as Clarke writes:

> "The rocks release their fossils,
>
> Their pink and white boulders, at the touch
>
> Of the sea. The slag the rivers
>
> Bring from the heartlands, works among

---

4 As Peach argues, "If it were called 'August Storm' the force of the poem would be lost. The 'Storm Awst' lives in the Welsh rural folklore and is peculiar to Wales" (1993:84).

5 And perhaps, as was alluded to in 'The City', the Welsh have a degree of complicity in this process through mining its land for 'foreign' investment.

The ancient stones. The new blocks stare
To England, above the blue surface
Where fine reflected weather smoothes
Over the estuary's silt
Of inherited sadness. But
I can hear it when the tide breathes
In, and spits the grit out on the shore." (Clarke, 1978)

Coincidence, Connection and Conclusion

As Clarke writes, Wales is a small place, a country of "coincidence and connection" (Clarke, 2008:9). Wales is, "small enough for us all to see the hills from the city, and to take a trip to town from the countryside" (PC). As a "small country" (Clarke, 2008:9), Wales is also vulnerable. In a century of capital flows which undermine traditional rooted communities, Clarke's plot- seeks to redraw the - lines which connect the Welsh to their landscape.

In this way, Clarke's writing has resonance with the nature writing tradition of North America. Clarke's geography is not a 'psychogeography' (see Chapters 1 and 12) – which could be interpreted as a temporary and itinerant walk around a place – rather it is a more long-standing 'incoming' and 'outgoing' (see Chapter 3), a regular tidal sweeping which accretes, abrades, deposits and washes meanings from a querencia into literary acts. Clarke's life and geography, her life(and)scape, is thus part of her writing; as she puts it:

> "someone once told me that I am my poetry, that all I am is there. I think and observe through poetry. Everything I have ever known, experienced, seen, heard, read is what I am." (PC)

Clarke's work thus translates her lifescape into literature. If 'nothing is until it has a word', Clarke puts new words to the taken-for-granted geographical belonging that defines the nation. Clarke 're-inhabits' a locale in Wales in order to re-sensitise herself to the rhythms of non-human life. Her poetry and prose functions like Snyder's *Axe*

Handles[6] (1983), she looks from her own present, and draws on her own past, in order to provide a template for the next generation's future. Clarke's vision is thus modelled on a nation which values its past ("The landscape where I grew up, its name, history and mythology, were Welsh" (2008:58)), but acknowledges how it's present is thoroughly entangled with its neighbours ("Two sea-sounds, two languages, two versions of the world, of myself, of history, of what should be valued, of what culture and civilisation are: English, the mother always sure she's right and knows what's best; Welsh, the father, luring to adventures beyond certainty, inclining me to the greater world. When we are most true to ourselves we are most universal" (2008:58)). In a moment where globalisation and cultural homogenisation are key threats (see also Davies, Chapter 10), Clarke's writings suggest that with the correct sensitivity to the

---

6 …
A broken-off axe handle behind the door
Is long enough for a hatchet,
We cut it to length and take it
With the hatchet head
And working hatchet, to the wood block.
There I begin to shape the old handle
With the hatchet, and the phrase
First learned from Ezra Pound
Rings in my ears!
"When making an axe handle
the pattern is not far off."
And I say this to Kai
"Look: We'll shape the handle
By checking the handle
Of the axe we cut with-"
And he sees. And I hear it again:
It's in Lu Ji's Wen Fu, fourth century
A.D. "Essay on Literature"-in the
Preface: "In making the handle Of an axe
By cutting wood with an axe
The model is indeed near at hand.-
…
And I see: Pound was an axe,
Chen was an axe, I am an axe
And my son a handle, soon
To be shaping again, model
And tool, craft of culture,
How we go on.

present and past, the future may be a bold and positive place to be; as Clarke articulates when placed at the meeting place between past and future (see Chapter 2), in Cardiff Bay:

> "I'm sipping a glass of chilled white wine at a pavement table outside Braz, the Millennium Centre restaurant, reflecting on the day, and watching people.
>
> ...a glimpse of the shining waters of the new lagoon, the headland at Penarth. The sea, held back by the barrage, used to come and go here. Under those calm waters are hidden limbs of mud which, before the Docks became the Bay, were resculpted twice a day by the Taff and the Ely and the huge tidal rise and fall in the Severn estuary. I loved those flooding and draining rivers, the mud, the half-submerged boats and buoys and shopping trolleys, but I enjoy this gleaming surface too. Down there, I know, the secret rivers still flow, arteries in the sea.
>
> I remember a hot childhood day with my father when we came to see the big ships in the dock. We walked between one dock and another, stepping over rails and railway sleepers where the trucks went, and just here, fluttering in the salty sea-grasses, we saw a small blue butterfly. It flutters still, among the impressions, the then, the now, and all that has happened in between, just here.
>
> It feels like September in France, evening in a sunlit square in a European city, a cafe table, a notebook, a glass of wine, and I'm glad of it." (2008:139/140)

Chapter 10
Durability and change: eternity and belonging in the plots of Grahame Davies

> "I think that all the lasting things are grey: the clouds above the mountains when it's late. When all around you changes, these things stay." ('Grey', Davies, 2005: 46)

> "We want everything which has a value to be eternal." (Weil, 2002:107)

> "People not things. People not things. That's what matters in eternity." (words from a drunk, companionless passenger, on 'No. 62 Bus', Davies, 2005:47)

These opening quotations illustrate the key themes explored in Grahame Davies' novel *Everything Must Change* (2007). Originally published in Welsh (titled *Rhaid i Bopeth Newid* (2004), the novel explores the moral geographies of time, durability, and change, and their aggregate effects on language, culture, and identity. In short, the novel explores ideas of belonging, at both a local and national scale. Davies argues that the issues associated with belonging are particularly relevant for a Wales in the 21st century and the changes the country is facing. Like many regions and nations, Wales is facing the threats associated with late modernity (see Beck, 1992, 1994, 1995); as Davies notes in the narrative, these threats constitute a "powerful, mobile and international surge of self-justifying and unstoppable deracination. [They] were difficult to define, let alone name, [but] Globalisation was the closest anyone could get" (2005:196). Globalisation, led by the mobility and ubiquity of corporate capital, but also driven by dominant nation states, places many threats on the uniqueness and identity of people and places. The familiar argument follows that as trans-national capital seeks to spread and make profits across the globe, everywhere becomes incorporated within the capitalist system, structured around the profit principle, and in this process places lose their individuality (see Ritzer, 2002). As a consequence, once taken-for-granted ideas of identity, language, and belonging are placed in tension as colonising ideas are imposed.

Davies identifies these tensions and their effects, and in order to explore them, draws historical parallels with other cultures that have faced threats to their borders as the world begins to change; as he states:

> "what I'm trying to do ... is overcome stereotypes; it sounds like a bit of a simplistic project but I am fascinated by that whole business … how you can view others whom you've been taught to suspect or be sceptical of – how you can come to a different view of them because I think within that potential there is the ability to do life-giving things." (Davies, Personal Communication (PC))

Davies is not an apologist for empire building or globalisation; rather he is interested in how people respond to these processes which affect the fundamental relations between people and place. He seeks to explore how we do, and how we should, retain a sense of who we are and where we are, when all our taken-for-granted assumptions of belonging, birthright, ownership and identity are under challenge. More accurately, he is interested in how we even consider this change – where it comes from, what it seeks to do, and how our geographies will be transformed as a consequence. Do we, for example, configure an inside / outside binary when understanding change, with boundaries constructed and defended from within? Is this change wrought from 'outside' the borders of our places by 'them', 'there', affecting 'us', here'? Or do we consider change to be a 'natural' and inevitable process, with our identities, culture and sense of belonging mutable and changing, "like the retelling of a story" (Williams, 2002:190)? If so, how do we retain a sense of ownership over this change and where it may take us? Can we retain a collective sense of 'value' of something 'eternal' that unites us in this new world? At the same time, how can we evolve to the changing world if we are tied too closely to the past? Davies deals with these questions through interweaving two key stories, alongside their histories and geographies, in *Everything Must Change*.

I talk to Grahame Davies about these issues in an office in Cardiff. From here we reflect over issues of belonging, Welshness, and language; our conversation follows in the footsteps of his characters, traversing from Cardiff Bay, through the urban city, into the valleys, and up to the north of the country. It is a good vantage point from which to take in, but also to reflect on, the changing nature of Welsh belonging in the twenty-first century. In his novel, Davies engages with these issues through entangling fiction and reality, explicitly so, in mixing the stories of Simone Weil – a French Jew and political activist of the twentieth century – and Meinwen Jones – a fictional Welsh language activist in the turn-of-the-millennium Wales. He describes how Meinwen's radical commitment to the Welsh language is a product of a particular type of Welsh identity; as he states:

> "the traditional ways of creating identity in Wales have been birth, so you were born into a Welsh identity and so that is essentially a biological way of belonging. You speak the language, that's the other way of identifying with it and very often that has been something you were born into as well. So you've got birth and language and belonging and – possibly because they have lacked the external superstructure of a state[1] – they were felt more strongly than perhaps they should to the point where other people might look at them and think, hang on – what's with all the birth and language and belonging?" (PC)

Meinwen Jones's commitment to her locality in North Wales and to the nation as a whole is encapsulated by her defence of the Welsh language. Echoing hard-line protests in the 1970s and 1980s (see *Fires on Llŷn*, Clarke, 1998, and Chapter 9), Meinwen seeks to protect her ideas of Wales through ensuring her local places and language remain despite the incursive threats of modern society. Jones' experiences are paralleled to those of Weil, occurring more than half a century earlier, as Davies explains in the prologue to the novel:

> "The idea for this novel was conceived when I was writing a volume of literary criticism and cultural theory based on my doctoral work in the Welsh department of the University of Wales, Cardiff, which included a study of Simone Weil. I was struck by the way Weil's remarkable life story would lend itself to a creative narrative treatment. Therefore, a few years later, I began a novelisation, combining her story with that of a fictional present-day activist, so that issues of conscience and political engagement could be explored in a contemporary context. The book is based on Weil's life story, and nearly all the major characters are based on real people, although telescoping, simplification and compositing have been necessary in both narrative and characterisation. ... In the story of Meinwen Jones,

---

1 In 1997 Wales held a referendum on the subject of creating a civic entity with devolved governing powers from the UK. As a consequence of a successful 'Yes' vote, the Welsh (Assembly) Government was established in 1999. Prior to this, Wales was wholly ruled by government in London, England.

> all contemporary characters and situations are imaginary, and any resemblance to real people is accidental." (Davies, 2007:i)

In the 'telescoping, simplification and compos[i]ting' of Simone's story we see what Collins referred to as the inevitable sifting and revision of a narrative, even a 'real' one, in order to gain a truth for the author and reader (see Chapter 7). In this subsequent entanglement of Simone's and Meinwen's plotlines we gain perspective and comparison on issues such as colonisation and invasion at a range of scales (the national, regional, local and personal[2]). We also see explicitly how in this case, the real and the fictional engage one another and entangle together to form the plotline of the novel: in literary geographies there is only page in place, and place in page.

"So I've been always been fascinated with these border experiences, [with] people who encounter and cross borders" (Davies, PC)

I begin by asking Grahame Davies why he chose to engage with issues of colonisation and its effect on belonging for the plotline of *Everything Must Change*:

> "It comes from the fact that I was born and brought up in a village which is literally on Offa's Dyke. I don't mean the current political border of Wales which lies six miles to the east of my village, but the actual earthwork which physically runs past one side of the village; so consequently I have always been acutely conscious of being born on the geographical border. Look east [from my home village] and you see an unbroken panorama as far as the Derbyshire Peak District. Look west and you see quarter of a mile to a brown wall of mountain which is the beginning of the Welsh Mountains, so physically [and] geologically it's on the border. [It's is on the] linguistic... border [too]: we were Welsh speaking, we did Welsh things, but we looked towards England. Our English accent was heavily accented by Scouse, we could be mistaken for Scouse even by Scousers and certainly by other Welsh people. So the sense of the marginal and the liminal has always haunted me and consequently given me a fascination with people who actively engage with it in one way or another and who cross it or challenge it in some way or another. So that I think where that came from and what's going on in "Everything Must Change." (PC)

---

2 As well as in terms of the "the biggest assimilation process of any minority in Europe both in the second world war, and since" (Davies, 2007:34).

This interest in border experiences is made manifest in the novel through Weil and Jones identifying the importance of being rooted, both geographically and politically, on one side of a border. Both seek to defend place – and their vision of social identity and geographical location – from the threats perceived to emanate from the other side. Due to the affective relation between people and place, for Weil and Jones the constancy and stability of place is deemed a common good; as Tuan tells us: "by remaining the same, [our places] tell us that, for all the loss of hair and accretion of weight that make us almost unrecognisable to ourselves in the mirror, our basic values and selves remain more or less intact" (2004:49). One feels that Weil and Jones would agree with Tuan as he elaborates on this point in the following way:

> "our home and hometown should stay the same. Now, why 'should'? How come a moral command has slipped into the language? ...[because] our own sense of self depends on such stability. When the neighbourhood we grew up in is demolished, we feel as though a part of our own personality is undone;
>
> and this would still be true even when the old neighbourhood has been improved. ... Any number of sound ecological reasons exist for preservation, but the reason I wish to emphasise here is the human need for stability – for a place where time seems to stop, a place to which we can return, a place whose wholeness and integrity confirm our own. (2004, 45/47–48)

As I have outlined in Chapter 1, as a consequence of these roots, individuals have a sense of place which they see as being bordered and defensible. This approach to place considers sites to be static in the sense of being changeless, preserved, and constant. Places are created over long timescales, with stable and durable place identities which are (and should be) widely recognised and acknowledged. Davies is sensitive to this human urge to defend our place and sought to explore its consequences through the book:

> "I do understand that desire to retreat and retrench and defend and otherwise keep and conserve what you value. That's completely understandable human behaviour. Just about every society of every conceivable kind has that inbuilt defence mechanism and TS Eliot said it is the instinct of every living thing to persist in its own being and so anything that's alive anything that has any kind of cooperative life will want to perpetuate itself very often by defensive means..." (PC)

In *The Need for Roots*, Simone Weil articulated similar sentiments:

> "To be rooted is perhaps the most important and least recognised need of the human soul. It is one of the hardest to define. A human being has roots by virtue of his real, active and natural participation in the life of the community, which preserves in living shape certain particular expectations for the future. This participation is a natural one in the sense that it is automatically brought about by place, conditions of birth, profession and social surroundings. Every human being needs to have multiple roots. It is necessary for him to draw well-nigh the whole of his moral, intellectual and spiritual life by way of the environment of which he forms a part." (Weil, 1995:53)

As Relph argues, "to have roots in a place is to have a secure point from which to look out on the world, a firm grasp of one's own position in the order of things, and a significant spiritual and psychological attachment to somewhere in particular" (1976:39). The fictional protagonist Meinwen Jones has her own version of this attachment and value system – as Davies suggests, "All her life, Meinwen had hated the idea of anything coming to an end: farms, chapels, businesses, lives, languages, anything" (PC), and this protection of the existing order was oriented specifically towards Gwynedd and North Wales.

In the passages of the novel set in the north of Wales, the reader is struck by the grounded and affective descriptions of the geography and community of the area, an area that is clearly resonant with meaning for Jones, as the following excerpt set in Porthmadog outlines:

> "Sunrise through her uncurtained window; sunset on the estuary below Bryn Aur. The scent of headier air when she stepped out of Yr Hafan in the morning, and in the evening, the sound of the wind under the door. The warm spring sun on her face; the rough wool of her frayed old jumper. The familiar solicitude in the voices of friends and family, and the unexpected friendliness of strangers. The minutiae of her neighbours' lives chronicled in the weekly paper; the serious, patient company of her books. Watching the street from a pavement cafe; watching the fields from the window of a bus. The thoughts prompted by every corner of the landscape she loved; the memories so intertwined that she no longer knew which were her own and which were other people's, their stories all interwoven into one web of belonging." (Davies, 2007:264)

In North Wales, a place of family history, personal memory, and local attachment, Meinwen feels at home. This attachment, allied to her use of the Welsh language, ties her into this place through politics as well as everyday conversation. For example, when visiting a café in Colwyn Bay (a location to which Meinwen's parents had retired), Jones is rooted simply by the local cadence in her spoken English:

> "He recognised the tell-tale signs in her accent.
>
> '*0 ble chi'n dod 'te*' he asked. 'Where do you come from then?' It was the traditional greeting of Welsh speakers everywhere. Profession, interests, qualifications, social standing counted for little to them. All that mattered was people's community of origin. The geographical establisher was the preface to a benign interrogation that would continue until the interlocutor had found at least one contact, however tenuous, in common: a mutual friend, a schoolteacher, councillor, chapel minister. Once that link had been established, they could relax." (Davies, 2007:72)

In North Wales, and Gwynedd specifically, Meinwen was very much part of a local community whose type was identified and valorised by Weil and Tuan (above); it is this community that bestows to her a sense of identity and "a firm grasp of one's own position in the order of things".[3] In this respect Jones' plotline draws on Davies' own childhood experiences:

> "Yes, [Meinwen is] absolutely [rooted in North Wales]. I mean those memories were my own of course, borrowed from my home village in north east Wales and my childhood holidays in Colwyn Bay. I also put Meinwen [in] the real café I went to, the 'Eat Easy' café in Colwyn Bay as it's still there and where I used to go as a small child. It was on a revisit there that I had the experiences that I then gave to Meinwen so it's a real place, so they are my memories just adapted [from] where I used to go on childhood holidays." (PC)

---

3 As Davies elaborates in correspondence, "Meinwen's identity is very firmly in Gwynedd in north west Wales. I gave her a childhood in Caernarfon because that is the urban area with the highest proportion of Welsh speakers, and is in the absolute heartland of Welsh-speaking nationalist Wales. Porthmadog, which is also in Gwynedd in north west Wales, has a similar proportion of Welsh speakers. By childhood upbringing (Caernarfon) and by adult choice (Porthmadog) therefore, Meinwen is a product of the Welshest places that can be found, with 86% Welsh speakers and 75.9% respectively (2001 census)".

Figure 10.1 Colwyn Bay. (Source: Author)

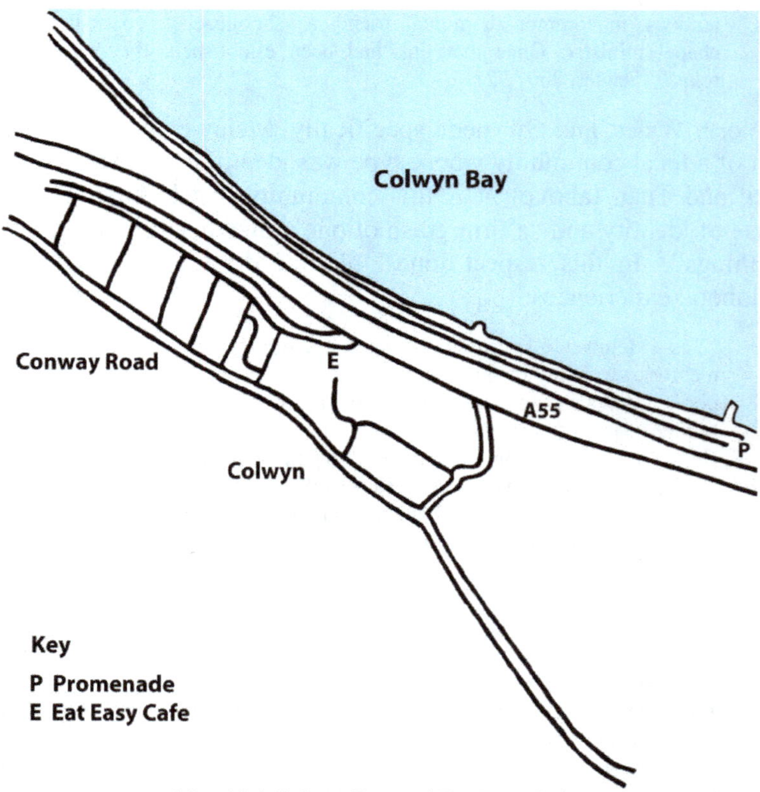

Map 10.1 Colwyn Bay and Eat Easy Café.

This rooted association with North Wales gives Meinwen her sense of hearth and hiraeth, her heart and politics were, both literally and metaphorically, always “set away, heading north” (Davies, 2007:67). For her, the south of Wales, and Cardiff in particular, was the antithesis of the rooted, grounded north. A location that in her eyes has been undermined by in-migration and colonisation from England meant the erosion of a conserved and stable sense of community that was anathema to her politics. As consequence, Meinwen didn’t have (and didn’t want to have) any face to face communications with people in Cardiff, for her it was an empty space to move through and a space to be apart from, not a part of. As readers we see her abstracted from the city, moving through it in a taxi, as if the car’s windows were a prism of her political views:

> “The taxi passed scores of other premises named after places and people which had nothing to do with Wales: St Kilda, Holme Lea, The Lincoln Hotel. Meinwen habitually scanned the written evidence of her environment for signs of change, growth or decline in the language’s fortunes. Every street was a fever chart of her sick nation.
>
> Renaming was one of the most galling aspects of the ‘colonisation,’ as the Movement called [it]. ... Meinwen thought of the region where she had been brought up. Incomers were already the majority in many of the more scenic parts; the remaining Welsh speakers being regarded by many of them as surly old rustics somehow still impertinently hanging around....” (Davies, 2007:34)

Figure 10.2 The Animal Wall, Cardiff Castle. (Source: Author)

> "Over Cardiff bridge, past the Animal Wall, past the entrance to Womanby Street, where clubbers would be hurrying out of the rain on their way to Clwb Ifor Bach, the Welsh-language nightclub where Dewi had planned for them to go tonight if Meinwen had felt up to it [see also Chapter 3]. ... But it was the cause, not a missed night out, that most occupied Meinwen's thoughts as she sat silently beside Dewi, looking out of the taxi window at the drenched city.
>
> The rainwater fed the swollen river Taff over which they had just passed; it pooled on the corner near the Angel Hotel, more than the drains could cope with, flowing up over the pavements, spreading the debris of the city's gutters ever wider. The rising waters brought the image of Cwm Tryweryn to Meinwen's mind like a recurring dream." (Davies, 2007:34)

As we will see, the image of the rising waters of the Tryweryn are not simply a poetic metaphor employed by Davies, but rather seek to draw political parallels between the imposed flooding of communities in North Wales by the British Government in 1965 in order to provide Liverpool with water, and the pollution of Cardiff by English language and culture. The Tryweryn flooding was an event that still haunted Meinwen:

> "She had not been born when Tryweryn was seized, its people evicted, their homes and chapel demolished and the valley flooded to provide water for England, but she had absorbed the black-and-white image of the desecrated community into herself, and felt the loss and exile like a personal pain". (Davies, 2007:34)

We will return to the metaphor of water below, but I ask Davies about the importance of this taxi ride in setting up Jones' alienation from the capital for the reader:

> "I think in some ways it is intended to show the limitation of the world through which Meinwen views things; it is primarily political, rather than as the way the actual inhabitants in the city [experience] it as 'the warm environment for human souls', as Simone would call it. This is how most people experience their localities, but anyone who has an obsession whether it's political or a cultural one will only see the world through those eyes. ... [For example] I love literature and I recognise this in myself and I view the world through the eyes of literature so you know if someone mentions a town I think who's written about it and that immediately becomes my way of viewing that town.... so Meinwen sees the world in that very specific way so that, her disconnect from Cardiff is largely intended to show that there are limitations in her strengths. Her strengths, you know that wonderful

> implacability and commitment she has, are also a weakness because she can't see the world in any other way, she can only see what speaks to the cause." (PC)

Here Davies notes how, for him, the world is always seen through the prism of literature, and thus resonantly echoes with the themes of *Page and Place*. However for Meinwen, the world – particularly the world of the capital – is seen through the prism of her politics, and the negative effects she sees caused by the influx of English and international culture on the community of Cardiff. This 'wonderfully implacable' commitment to her politics is laudable – she stoutly defends her place in the face of colonising forces – yet it nevertheless remains partial: it obscures her ability to see the different communities that exist in the capital. There are, indeed, other versions of the city beyond her 'polluted' view (as we have already evidenced in Finch (Chapter 2), Williams (Chapter 3), and Robson, Chapter 4)). These other versions of Wales, and their view of Meinwen's Welsh Language campaigns, are summed up in the novel by Cardiff-based newspaper journalist John Sayle:

> "God, they annoyed him. He damn well hated their self-righteousness, the implicit assumption, whatever they said to the contrary, that they were the real Welsh, that people like himself were some kind of mongrels. He wouldn't have these woollybacks making him feel like a second-class citizen in his own country. They thought their Wales was the only one – some humourless, inbred, slate-grey redoubt in north west Gwynedd; they had no idea of the warmth and value of communities like the ones where he'd grown up, with their belonging, their socialism, their shared wit.... He knew the value of the real Wales: it was epitomised by the Valleys of his youth, and he'd defend it against all comers." (Davies, 2007:45)

Here Sayle articulates one perspective that is obscured from Meinwen's view: that there are many communities within Wales, perhaps many versions of Wales, and that each person who is part of those communities feel that theirs is a real, authentic, and valid definition of the nation. From Butetown, with "the remnants of the traditional multi-racial community, Yemenis, Somalis, West Indians" (Davies, 2007:134), to the valleys defined by waves of immigration in the nineteenth century and "now as Welsh as they come" (Davies, 2007:125), Davies writes a nation that is multi-vocal in its voice, rather than singular in nature. This acknowledgement within the text, if not by Meinwen Jones, raises questions that are at the heart of the novel: is Wales a singular entity that can be united under one idea of belonging, or is it an agglomeration of regions which all identify with their localities in different ways? One may argue that Meinwen's close affinity with Gwynedd and its surroundings represents the core of her being, language, and identity. However, as we have seen, in other areas of Wales competing ideas of what it is to belong are created which may not be compatible with such stable, rooted notions. The entanglement of Simone Weil's story, using the example of World War II and Germany's invasion of France, offers a useful insight into this situation:

> "'The fall of France has focused my mind as never before on the history of the Languedoc as a region...' [said Simone's father].
>
> Simone, speaking for the first time that evening, interrupted him. 'Once, it wasn't a region, though, was it? It was a nation, with its own language and culture.'
>
> ... 'French patriotism is so arbitrary,' Simone continued, getting into the stride of her argument. 'If you think about it, it's just an equation between absolute good and a particular territorial area. If you dare to question the territorial element of the equation, they call you a traitor. Take people in Provence and Brittany – and Alsace.' She looked at her father, who looked away, embarrassed. 'We expect them to forget their past and collaborate with France even though most of them have a different ethnic or linguistic heritage, but then we condemn anyone who collaborates with Germany, which only wants to do to us what we do to our own colonies. There's no logic to it'." (Davies, 2007:187)

Here Weil draws attention to the fact that the France we know today as a unified nation is in fact an agglomeration of smaller 'nations' – of local areas with their own 'ethnic or linguistic heritage' which have unified together under threat of a common enemy to form a new nation. If all nations are formed in this way, she argues, the process of nation building is not natural or inevitable, but rather arbitrary and political. National belonging cannot be created in the same way as local belonging as the processes which form small-scale attachment – community knowledge, face to face communication, and everyday lives lived in close proximity to people and place – cannot be recreated at a larger scale. It is impossible to know, meet, and talk to all the citizens at a national scale in the same way as it is in a village or neighbourhood. According to Benedict Anderson (2006), national unity and fraternity therefore needs to be 'imagined' rather than directly lived. From this insight we can identify how Meinwen is engaged in a process of 'scaling up' her own local connections and imagining a national community of Wales. As Davies puts it, this is "the way that most people experience their Welshness; for most people their experience of their country is often an unquestioned immersion in their locality" (PC). However, for Jones, this unquestioned entanglement of Wales into her locality, and vice versa, means that the threat posed by the spread of English language and international culture is a direct threat to the way in which she imagines her community: she realises for example that,

> "soon... there'd be no Welsh-speaking communities left to save; one more irreplaceable cultural ecosystem would be extinguished, the one which had created her, loved her, given her a sense of belonging, and without which her life would lose all meaning. The words of her earliest memories, the words of the poetry she loved, the words which had named the fields, the rivers and the hills among which she was raised, and which bound all those tilings and more together in one living, intimate web of belonging, they would become silent forever." (Davies, 2007:37)

Jones' imagining of her local onto the scale of the nation is thus a political and a moral choice, she wishes to protect what for her is the essence of her own identity – her place and her language – and she imagines this identity to be 'Welsh'. For Meinwen, the Welsh language remains an important key to unlocking the identity of Wales (see Box 10.1). The defence of this language becomes a touchstone for defending her place and history, as the following scene goes some way to illustrate; here Meinwen confronts an English incomer, Connor, who has recently bought empty Hermon chapel next to the isolated former manse, Yr Hafan ('The Haven') where she lives:

> "'Hi, Manwen.' After knowing her for three months, he still hadn't bothered to learn how to pronounce her name properly.
>
> But that wasn't Meinwen's main concern just now. As the dust cleared, she saw what Connor had been doing. He'd been using an angle grinder to remove the carved wording from the foundation stones at the base of the chapel. Where previously, the giants of the cause who had laboured and sacrificed to build Hermon had been commemorated in sandstone lettering, now there was only bare scraped stone. The cloud of witnesses had become a cloud of dust. She stared at the now wordless masonry.
>
> 'What have you done?!'

> Connor straightened up. 'Oh, those old inscriptions,' he said. 'Bad vibes, you know? We needed a new start. I'm going to carve some symbols there instead.' It was clear he hadn't the faintest idea of the significance of his actions.
>
> 'But, those were the people who built this chapel!' Meinwen could hardly find the words. 'Those people have relatives still living round here. That was our history!'
>
> 'Yeah, but it's our house now, and, you know, a new start. A clean slate, like.'
>
> Meinwen was numb. She couldn't begin to try to argue. She almost ran back to Yr Hafan.
>
> She threw herself down on the sofa, and hugged her knees to her chest. Not even the words on the stones. They won't even leave the words on the stones. We're being scoured out of history while we're still alive. It's like a nightmare. Not even Yr Hafan was a haven any more. She had reached the end.
>
> She thought of things she could say to Connor. The argument chased itself round her mind, trying to find the words that would make the noise stop, make things be as they were, stop them changing. The argument was in English, the words finding ever more familiar routes through her brain. Not even her mind was safe. English was pushing into it, colonising it cell by cell, possessing its neural pathways, Saxonising its synapses, supplanting, renaming, uprooting.
>
> Outside, the angle grinder started up again, the sound cutting through the air, through the walls of Yr Hafan, through Meinwen's head. She put her hands over her ears and wept." (Davies, 2007:236)

In this scene we see the geology of her landscape, once manufactured into a secure haven, now being re-scripted – the slate literally being wiped clean – by incoming forces across the border. Due to the umbilical connections between herself and place, this change to the landscape has an embodied affect, stifling her ability to think and speak (in Welsh) as it serves as an indication of the remodelled vision of nation and its culture she is witnessing. Davies comments on this scene as follows:

> "I was asked by a journalist once what that was based on and I said, 'well it was imaginary'. I've never even heard of that happening but it is symbolic of a kind of mind, a deracinating mindset, which really doesn't have any interest in the locality it was just a shorthand way of showing that. Meinwen is a product of her culture, it is what makes

> her what she is, her primary if not only value is her position as a member of the Welsh speaking community, so anything that diminishes that diminishes her." (PC)[4]

Davies finds this territorialisation of morality and language compelling as it forces us to decide which side (of the border) we're on:

> "... fundamentally I am completely fascinated with moral choices, I find moral choices thrilling as people make them unconsciously many times every day... so I'm very interested in diving into the consciences of the characters and putting them into situations where they have to make decision of conscience." (PC)

Due to the overwhelming sense of personal attack Meinwen experienced following the changes to her locality she felt she was left with no moral choice but to break the law to draw attention to the issue. Meinwen committed criminal damage by breaking a local estate agent's window, and was subsequently arrested and charged, as the novel reports:

> "She got three months. Her string of previous convictions allowed that length of sentence, and the worryingly destructive nature of her protest had led the magistrates, who were not entirely unsympathetic to her aims, to decide that they needed to show that her actions on this occasion could not be condoned." (Davies, 2007:242)

In line with Gillian Clarke's *Fires on Llŷn* (1998), Meinwen was moved to political action as a consequence of the threat she felt to her constitutive coingredience with her local place. However this radical route to protect herself and sense of nation could be interpreted in another way. In the same way that her view of Cardiff wiped clean the communities that drew succour from that place (and the creation of a unified France had the effect of silencing regional identities in Weil's

---

4 A similar scene is depicted in Gillian Clarke's poem 'Border' (1998:21):

Border
It crumbles,
where the land forgets its game arid I'm foreign in my own country. ...
At the garage they're polite. "Sorry love, no Welsh." At the shop I am slapped by her hard "What!" They came for the beauty but could not hear it speak.

story), her radicalisation could be read as an attempt to silence the right of others to live peacefully in that area of Wales. Through a media debate between Meinwen and Sayle, Davies focuses attention to the heart of the argument:

> "Meinwen was continuing.
>
> 'If you look at Denmark, they have a property law preventing their houses being bought up by Germans. So does Poland, for the same reason – because their economy isn't as strong as Germany's. Laws like this are normal throughout Europe. It's Britain that's the exception. By refusing to have such a law here, you're letting the last communities on earth where Welsh is spoken be exposed to the full force of the uncontrolled market. Which means that Welsh is going to be wiped out as a community language just so some rich people can escape the cities and enjoy a nicer view. A Property Act is a question of simple natural justice...'
>
> 'Simple natural racism.' Sayle hadn't even waited for his cue. He spoke with dismissive authority, as if the collective impatience of a nation was impelling every word. Years of conducting the public soliloquy of a newspaper column had given him the apparent conviction that his views were shared by all but fools, whom he did not suffer gladly.
>
> 'Let's be clear,' he went on, 'this Property Act business is just a front for your childish hatred of all things English, for your primitive rejection of cultural diversity, for your backward-looking, fascistic, bigoted mindset which is bent on forcing the Welsh language down everyone else's throats.
>
> 'There is no place, no place at all, in a civilised democracy, for the kind of ethnic cleansing you're proposing. We all know the results of this kind of thinking – we've seen it in Nazi Germany, in Kosovo, in Rwanda, in Darfur'." (Davies, 2007:25)

In comparing Meinwen's position to the example of Nazi Germany we not only see the merger of Weil's and Jones' stories but also competing visions of national identity. Although the comparison to Nazi Germany is utilised by the author and the character John Sayle for tabloid-style controversial effect,[5] it does raise the key point that in

---

5 In 'Everything Must Change', Davies makes the point that "all the characters are good, there's scarcely any villains in it" (PC), in other words, everyone is motivated by their own moral convictions of right and wrong. However, the inevitable consequences of moral acts are the creation of winners and losers, and each of us have to determine where we stand on each action.

attempting to defend an idea(l), one also has the effect of marginalising alternatives. For Meinwen, it was crucial that the Welsh language lived, and this was more important than allowing other voices in Wales to have their say. However, as the novel enters its final stages and Meinwen realises the personal and political costs her entrenched position involves, her view becomes subject to change. As the following illustration may imply:

> "Meinwen was surprised that this man's English should sound so scouse, but his Welsh be so good. She thought about his question for a moment." (Davies, 2007:72)

Through the events of the novel, Meinwen Jones begins to see a political alternative to her committed resistance to change. As alluded to above, in everyday situations (such as an encounter with an indigenous Welshman in a café who spoke Welsh perfectly, but in a Liverpudlian accent) Meinwen's view is challenged by ideas which offer new visions of Wales; as Davies states:

> "Meinwen inhabits a Welsh speaking world which is almost hermetically sealed as a value system where everybody is radical, everybody is a pacifist, everybody is republican, everybody signs up to a particular political agenda; and her journey is to see if there are others ways of being Welsh, other ways of promoting the cultural survival [and belonging] which is at the heart of her project." (PC)

Inspired through a dream which changed the way she saw the symbolism of water, Meinwen came to see the influx of English language and culture not as something that eroded and drowned valuable things, but as something which could be swum through, and even could be life-giving; as Davies recounts:

> "At the end of the novel the crucial passage is where Meinwen has a dream which subverts the traditional nationalist interpretation of the imagery of water ... water has always associated with inundation, with drowning, with suffocation, with oppression, with any number of bad things, and it is a shorthand which simply doesn't exist in a culture which hasn't had that experience. [Meinwen has] been literally immersed in that world view from birth, but she is now enabled to see water as something else: she can see it as fluidity, as movement, as flexibility, as adaptation, all this image which she has been raised to believe can only be negative is eventually redeemed and she has the opportunity to see it in another way." (PC)

The consequence of this new interpretation was significant for Jones; as the novel outlines:

> "The realisation hit Meinwen like a blow... Everything must change.... . Only now, when it was too late, did she understand what that phrase had meant.
>
> It didn't mean she should accept her personal destruction fatalistically; neither did it mean the Movement's belief that circumstances could be forced to fit their ideology, against all the evidence; neither did it mean standing like a sandcastle, insisting that it would be the waves that would have to yield in the end. No, it meant recognising that mutability was inevitable; it meant riding the waves of transformation, working with the forces of change, channelling them to directions consistent with survival, and changing with them when necessary. Immutability could only result in death. Being flexible, adapting oneself, that was how to survive." (Davies, 2007:265/6)

Through this re-interpretation of her position Meinwen began to create an identity that sought allies and common ground, rather than antagonism and isolation;[6] as Davies puts it:

> "[her new position] can be seen as a symbol of the kind of energies which can promote the same ideal of cultural survival but through flexibility, through charm, through influence, and through all kinds of movement, rather than opposing some or trying to maintain some immobile frontier against an external threat." (PC)

---

6 Up until this point, Meinwen had revelled in her personal, and increasingly political, isolation, as the novel outlines:

"There was something agreeable about the cold, though, she thought. The fact that you could feel it, that you could hear the wind roaring outside, it made you appreciate your comfort all the more. It felt a bit like the eternal Welsh experience: sheltering from external threat, cultivating an inner warmth against a coldly hostile world. It was not such a bad state. It kept you awake, that was for sure. Perhaps the soul needed an external inclemency in order to prove itself". (Davies, 2007:93)

Box 10.1. Language and Identity in Wales

The story of Meinwen Jones draws attention to the role of the Welsh language (Cymraeg) in the history of the country. In the past, the language has been undermined and culturally colonised by the English and their lingua franca. In the nineteenth century, for example, the language was deemed uncivilised, as the following Editorial from 'The Times' in the 1890s suggests:

> "The Welsh language is the curse of Wales. Its prevalence and the ignorance of the English have excluded, and even now exclude, the Welsh people from civilisation, the improvement and the material prosperity of their English neighbours... their antiquated semi-barbarous language, in short, shrouds them in darkness." (Cited in Abley, 2003:255)

Elmes outlines how this view has roots even further back in history, as King Henry VIII's Acts of Union (1536-43) served to criminalise the use of Cymraeg in order to build a 'British' nation, as Elmes outlines, these Acts:

> "outlawed the language of Welsh... [their] intention [was to] 'utterly to extirpe alle and singular sinister usages and customs' [including language]: "The people of the same dominion have and do daily use a speche nothing like ne consonaut to the naturall mother tonge used within this Realme." (2005:78)

As Rob Penhallurick, Head of English at Swansea University sums up, the message on language was clear: "If you wanted to get on, you had to learn English, not Welsh" (Elmes, 2005:84).

In 1993, however, the Welsh Language Act was passed by the (British) Government which put Cymraeg on an equal footing with English in terms of the public sector in Wales. As a consequence, the mixing of English and Welsh is increasingly common in both formal and informal communication. As Elmes continues:

> "the English dialect of south Wales, despite being strongly English in its overall feel, possesses many words with Welsh origins.[As Penhallurick states:] "When people say 'there's lovely', again it's a translation from a Welsh idiom: the Welsh word is 'dyna' – there's – followed by an adjective. The other thing that Welsh English speakers are well known for is they tack 'isn't it' in to sentences: 'We went shopping the other day, isn't it', which again sounds odd. But once again it's a direct translation of an idiom from the Welsh language [and] because of the way Welsh verbs are constructed, its quite normal to have that sort of tag at the end of the phrase or a sentence. Then it get transported into Welsh English." (2005:86)

Despite the adoption of Cymraeg as an official language for Wales and a creole Welsh-English emerging in South Wales, there remains many communities in Wales that speak variations on their own form of Welsh. As Elmes states:

> "Take for example the story of Deborah, who's spent all her life in the Welsh capital – born in Roath and now living in the Splott district – about a trip she'd made to the north:
>
> "I did 'ave a negative response when I went into a shop and the lady just blanked me; so I asked 'er again what I wanted and she still blanked me; and in the end she did serve me. And we come out from the shop and we joined the people from the coach an' they were sayin' the same, and we said, 'Well, why?', an' it was because we were from Cardiff. I was shocked because [it was] Wales: we're all Welsh – I was really shocked. Apparently they didn't like that Cardiff was the capital and they did prejudge us by our accent to start with, but it was more boundaries. But I thought it was a bit strange they judged us by our accent first off, because we're just the same as them!" (2005:79)

Acknowledging the diversity of accents, dialects and variations of Welsh across the country suggests that the nation could be argued to be a loose assemblage of regions, rather than a united nation with a singular voice.

Conclusion. Liquid Lives: "It's written as a transitional process for Welsh identity" (Davies, PC)

We have seen in this chapter how Davies explores ideas of eternity, morality and belonging and their connections to the geographies around us. Echoing Bauman, Davies' characters "are haunted by the problem of identity" (Bauman, 2005:6), they are concerned with the values of the world they inherited and the world they are going to pass on. Throughout most of the novel, Meinwen Jones is a 'modern', she is wedded to the fixity of borders and the permanence of an identity anchored to these political and geographical roots. For Jones these connections are located in the Gwynedd area of North Wales, but she nevertheless assumes these ties are national in nature as she scales up her allegiance from her immersive locale to the level of Wales. However, the world in which she lives is changing, and Davies' book details the transitional consequences for Meinwen's identity at the personal, local and national level. Where initially Jones adopted an intransigent position, a 'sandcastle' stood fast in front of changing tides, the rise of devolution and language creolisation within the country realised a change in her position. Meinwen became a 'liquid modern' (after Bauman, 2005). As Bauman puts it: "Life in a liquid modern society cannot stand still" (2005:3), and as a consequence, it becomes necessary to quickly reflect, scrutinise, and adapt to new conditions. Liquid moderns realise the importance of reinventing belonging; "producing, staging, and cobbling together" (after Beck, 1994:13) their identity between shifting sands becomes not simply a choice, but a responsibility (after Bauman, 2002). In practice, Meinwen adopted a new position, framing her identity as fluid and life-giving, being precisely the 'liquid' modern advocated by Bauman himself.

It is this process of reinventing identity and questioning the borders (be they geographical or political) inherited from the past that motivated the author in writing this novel, as Davies puts it:

> "... what fascinates me is the opportunity for people to transcend the boundaries which they have either chosen for themselves in earlier periods of their lives or which they've been born into. I find that potential, that challenge, that change, as the most moving and profound experience so that's what's going on [here], putting characters in a contested landscape in which they can face crucial

> moral choices as to how they're going to live their lives and how they can, if they wish, coexist with other people in a creative way in which people enrich one another and in which this tense competition can become something more benign." (PC)

Reflecting the process of change in Wales itself, Davies outlines the possibility and necessity to be a liquid modern in the present day. Where once language and birth were key to the debate on nationalism, now the rise of the self-government means the opportunity for new forms of belonging to be generated, as Davies puts it:

> "Now … you have the potential for different models of belonging in which you don't need to say well I'm Welsh because of birth or because of so many generations of belonging because I speak the language. You are Welsh because you are here and that is enough. …There's an energy at play there which is it seems to me entirely positive which isn't using those defensive – understandably defensive – energies." (PC)

In this new context Davies identifies the possibility for a new 'cosmopolitan tolerance' (after Young, 2000) between liquid moderns on all sides of the debate – between the different communities who inhabit Cardiff, to the different regions that feel their own idiosyncratic affiliations make them authentically Welsh. From this entanglement of page and place we thus identify a tolerance which has the potential to create a 'differentiated solidarity' (ibid.) which may come to define the future of the Welsh nation.

Part IV

Borderlands

Chapter 11

Entangling Owen Sheers: 'a conversation of place and page over time'

> "There are places that speak,
> telling the stories of us and them.
>
> A village asleep, loaded with dream,
> an ocean, flicking its pages over the sand.
>
> Eventually we reply, a conversation
> of place and page over time,
>
> inscribing the map so that each,
> in turn, might hold the line." (Sheers, 2009: IX)

Owen Sheers is a poet, novelist and scriptwriter. Born in Fiji but brought up in Abergavenny, his writings evoke the entanglements between page and place in order to explore their importance for who we are. His sensitivity to the entanglements between literature and landscape was formed and developed through his upbringing in the countryside of south Wales; as he writes:

> "I fell in love with landscape long before I fell in love with poetry. As a child and a teenager I felt different when I was outside, it was as simple as that. Running on the hills above Abergavenny, walking through a field in the rain, facing up to the breakers on the Pembrokeshire coast or looking at the view from the broken spine of the Skirrid, the hill I woke to, framed in my bedroom window, every morning." (2008:172)

Figure 11.1 A view of Skirrid Hill, Monmouthshire. (Source: Author)

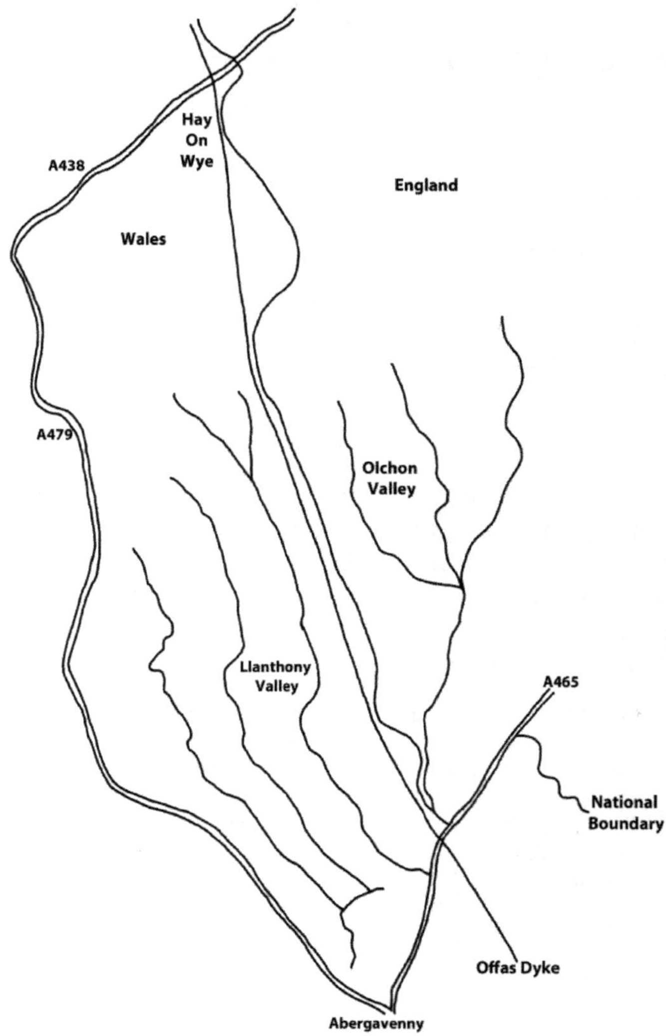

Map 11.1 The Llanthony, Olchon and surrounding valleys.

It was the countryside around the Skirrid that I walked with Sheers in order to discuss the influence of place on his writings.[1] As he outlined, these youthful experiences of the countryside moulded his connection to landscape, particularly the Llanthony Valley, Skirrid Hill, and Olchon Valley on the borders between Wales and England. The openness of the land mirrored his own approach to the places and communities of the area; as he states:

> "I didn't really come back to live here properly until I was 9, and actually perhaps ...that's a very happy situation for a novelist in that you're open to the strangeness of it as well, you can see it with fresh eyes." (PC)

Arriving in south Wales at a time when his outlook on life was beginning to form, the landscape provided for him a playground and an arena in which to develop. As he articulates in *Border Country*, for him and his best friend the open countryside was:

> "A place where we tested our voices,
>
> young as the buzzards above us
>
> striking their cries against a flint sky..." (Sheers, 2005: 10)

With the narrow valleys and broken hills so much a part of his everyday childhood, these landscapes formed Sheers' view on the importance of the 'ongoing conversation between page and place' which have come to define his writing. Through growing up here, Sheers lived and lost in this landscape, learning the lessons of the importance of roots, and our position as humans in the broader sweep of geological time. This is expressed in *Skirrid Hill*, an award-winning book of poetry that is split at its core by two entries: 'Y Gaer' and 'The Hill Fort'. Both poems outline perspectives of the death of his childhood friend and the role the land plays in dealing with life and consoling in tragedy.

> **Y Gaer (The Hill Fort)**
>
> ...
>
> so I think I understand why the man who lost his son

---

1 Owen Sheers was interviewed at the commencement of this project in Cardiff, but I walked more extensively with him on a tour of the Llanthony Valley arranged by Literature Wales.

comes here only in bad weather
when he can lean full tilt
against the wind's shoulder

take the rain's beatings, the hail's pepper-shot
and shout into the storm
finding at last, something huge-enough to blame

The Hill Fort (Y Gaer)
On a clear day he'd bring him here,
his young son, charging the hill
as wild as the long-maned ponies.
...
pointing out all the places lived in
by the fathers and sons before them:
Tretower, Raglan, Bredwardine...

and what he meant by this but never said, was
'Look. Look over this land and see how long
the line is before you – how in generations

we're no more than scattered grains;
that from here in this view, 9, 19 or 90 years
are much the same' (Sheers, 2005:20-21)

In Sheers' view, we both form and are formed by the landscape. As we find our voice in the places around us, we are influenced by the geology and cultures of that place. As a consequence, we position and define the land, and in turn the land positions and defines us. It is this constitutive coingredience (see Chapter 1) that gives the basis for our views on the world, the 'lines' we must 'hold' in order to root and secure our identity. It is these themes are that explored further in his novella, *The Gospel of Us*.

The Gospel of Us to *Resistance*

> "It was like my bampa said, wasn't it? Cleverest thing the Company ever did, to make us forget where we came from.... Well, right then, as the sun sank over Swansea, as the wet sand on the beach turned the colours of an oyster shell, he made us remember. And in doing so, made us see." (Sheers, 2012:45-46)

The Gospel of Us is Sheers' re-imagining of his play 'The Passion', which remade the streets of Port Talbot into an open air theatre in

Easter 2011. This event explored the importance of roots and belonging as 'The Teacher' (played by Michael Sheen) sought to remind the locals of the riches they possess in the entanglements of their history and geography. The theme of the book therefore echoes Sheers' interest in the role of page and place, but the form of the book is also noteworthy as it actively sought to build community in Port Talbot itself. As he writes in the preface to *The Gospel of Us*:

> "NTW [National Theatre Wales] published the text in three pamphlets, each released on the morning of that day's performance. As a writer this proved to be a unique experience. To sit in a pub and watch nine or ten people reading and talking about a fictional scene in which they themselves had just played a part. This was, though, typical of the nature of The Passion, and was always at the heart of what we'd hoped to create[:] a maelstrom of an experience in which the audience would be both witness and participant; three days of performance where ideas of place, history, fiction, theatre, life would all intersect within a single town." (Sheers, 2012:9-10)

The book was therefore not simply a story *about* place (see Box 11.1), but a story that *made* place; due to the entanglement between literature and location, and the power of stories to write a place into being, *The Gospel of Us* became an active agent in affirming the identity of Port Talbot and contributing to community building; as the young narrator of the story explains:

> "Whenever my Bampa was down the social telling a story, he'd pause before the best bits, lick his finger, touch it to his throat and then – so you could only just hear him — whisper, 'God'shonesttruth now.' Then he'd carry on, making all them other men lean in over their pints and cans, listening.
>
> Well, I'm licking my own finger now. And touching my throat. But it isn't God's Honest Truth what I'm going to tell you. It's ours." (2012:13/14)

Box 11.1 Page making place

In *The Gospel of Us*, Sheers' raises questions concerning what is important to people's lives in an age of mobility and change. Echoing themes in Clarke (Chapter 9) and Davies (Chapter 10), he asks questions regarding what defines us and what 'lines' we should 'hold' in such circumstances; in other words, he asks: what is it worth living, and perhaps dying, for? In the spirit of 'Passion' plays, which are often loosely constructed around the Easter story of the Christian religion, Port Talbot is woven into an entanglement of page and place through becoming threatened from ownership by the (fictional) corporation 'ICU'. The Company wants to displace the town's inhabitants in order to mine the geological resources in the area. In the face of this threat, The Teacher seeks to inspire and mobilise the community's defence, at no small cost. As the following excerpts, from the perspective of our young narrator, outline:

> "I looked up to the balcony of the Four Winds to where the Company Man had been watching with his lackeys from the Council. Then others turned to look at him too, until the whole crowd was staring at him, daring him to speak.
>
> When he did, it was with the voice of a beaten man. ... In killing the Teacher he'd showed everyone what the wealth under this town was.
>
> 'If that's what matters to you,' he finally called out, trying to muster some pride, 'then there's nothing of value here for us. We won't be back'." (Sheers, 2012:166)

In trying to explain the constitutive coingredience the towns people rediscovered as a consequence of The Teacher's actions, the narrator state

> "I suppose the most simple way to put it is this. Us, we came out of the ground. Our memories, our parents' memories, their parents' memories. And with them our stories. Everything that made us, everything that made the town more than just bricks and glass and concrete." (Sheers, 2012:165)

Here we see Sheers emphasising the role of stories in binding communities together, passing on inherited knowledges, entertainment and identity, and through so doing, creating ties between people and place. Echoing themes from Clarke (Chapter 9), Sheers outlines how these intangible resources are ultimately more valuable than whatever material resources can be mined in an area for economic reward.

The themes of identity, landscape, and insurgence against a colonising power come to define *The Gospel of Us*. These themes are further developed in Sheers' novel *Resistance* (2007), which was subsequently developed as a film (see Gupta, 2011). *Resistance* was set in the Olchon Valley, close to his childhood Skirrid Hill haunt. As we walked up this location, through Llanthony Valley, up and over Offas Dyke, and into the Olchon Valley, Sheers outlined how personal this location was for him:

> "the genesis of the novel [began] as I was thinking about trying to find a story that I could set against a landscape I knew, really knew much more intimately [than in my previous novels]. I mean I quite often say that [this] is a landscape that I felt I knew in the bone, in that level of bone marrow." (PC)

Set in the Olchon Valley, *Resistance* is a novel based on a counter factual history. As Piatti and Hurni (2009) tell us, a counter factual history, "is the description of a historical 'what if'", it is a speculative historical exercise that "revolve[s] around the basic premise that some event in the past did not occur as we know it did, and thus the present has changed" (Hellekson, 2001:2, also cited in Piatti and Hurni, 2009:33). As Piatti and Hurni point out, the moment at which history shifts to a parallel world is called a 'nexus event' (see 2009:33). In the case of *Resistance*, the nexus event is centred around the failure of British troops to take the Normandy beaches in mainland Europe on D-Day in the Second World War. As a consequence, in this counterfactual history German troops invaded Britain and made their way across England and Wales.

This nexus point brought the sweep of war across the country, and sought to wipe clean its culture and impose new national idea(l)s of how the world should be. In the remoteness of the Olchon, this threat initially seemed abstract and far-removed from the reality of everyday life:

> "They should have nothing to worry about here though, that's what Reverend Davies had told them. And the Home Guard officer who'd come round handing out the leaflets a week ago. 'Disable all vehicles so only you can use them. Hide food stores and essential supplies. Offer no resistance but offer no help either.' He'd said these sentences in a flat tone, their intonations worn thin through repetition. But then he'd given Maggie a quick smile and briefly found his own voice again. 'I wouldn't worry too much though, Mrs Jones. Really. There's no way Churchill'll let them past the beaches. And even if he did,

> well, to be honest, I doubt you'd see a Jerry up a valley like this'." (Sheers, 2007:45)

In Sheers' counterfactual history, such hope was misplaced. The novel begins when the threat to the valley's way of life is realised, and the women wake to the desertion of their husbands and sons. The men have left in order to mount an insurgency against the invaders, although the families they have left behind are necessarily unaware of their motivation for leaving. This insurgency was the point at which Sheers' counterfactual history entangles with real plans generated to form an underground resistance army in South Wales. This resistance army – called The Auxiliary Unit – was to be mobilised in the event of invasion, as Sheers explains as we stop and look down on Llanthony Priory from the foothills of Offas Dyke:

> "So having decided that I wanted to try and find a story that I could set here, I started to learn about this very secretive organisation called 'The Auxiliary Unit' which effectively would have been Britain's underground resistance in the event of occupation ... actually it's rather appropriate, since the first I heard about The Auxiliary Unit was when I was working for a builder in a garden in this valley, and we were working here through a long hot summer, and all sorts of stories were being told on that building site and one was that the farmers in this area had been given a cache of arms and if the Germans had invaded, they were expected to go into the village and begin their acts of sabotage." (PC)

As Sheers continues in print:

> "The Auxiliary Units were to be Churchill's last-ditch defence; a controversial and, in terms of international conventions, illegal government-funded civilian insurgency. Given the strategic importance of the area, and the possibility of an invasion via the Bristol Channel, there was a particular concentration of Auxiliary Unit bunkers in south Wales. In coalmines above Port Talbot, in the woods above Newport, on top of the Blorenge outside Abergavenny, around the village of Usk and, I'd heard rumoured, in the valleys south of Hay-on-Wye." (Sheers, 2011: no page)

Figure 11.2 Llanthony Valley and Llanthony Priory. (Source: Author)

Due to the enactment of these auxiliary units, in *Resistance* the women of the valley wake to find their husbands gone. In leaving to mount insurgency strikes, the men have left their farms and families unguarded as German patrols take control of the valley; as the novel explains:

> "now this wounded [valley] was ... exposed in the way an injured lion or bear is exposed. Vulnerable but rageful and thrashing too. Open to its enemy and to the very extremities of aggression as well. Albrecht knew [this]. ...It would be boiling at every turn, with resentment, with anger, with desperation. It would fall, of course; it seemed they all did in the end. Even Moscow and Stalingrad had fallen. But how many of them would it swallow first? How many of them would have to tip themselves into [its] jaws before the [country] finally choked on their blood running in the veins and arteries of her streets?" (Sheers, 2007:32)

As it's title suggests, *Resistance* is a novel about the moral issues surrounding the occupation of this valley. How would the invading troops, battle-hardened but war-weary, colonise this undefended valley? How would the women of the valley respond? Would they, as they had been told, 'offer no help', or would they openly resist?

Indeed would they be complicit with the enemy now their husbands had departed? Through an exploration of these issues, the novel asks key questions regarding what, where and whom we should live for? The novel focuses in particular on how these moral questions affect one housewife, Sarah, and her relations with the German patrol officer, Albrecht. As Sheers confirms:

> "the title *Resistance*, for me really applies to Sarah's resistance in relation to that 'love story'. ... It's a triangle of a love story really isn't it, 'cos I don't think Sarah ever falls out of love with [her husband] Tom, but here is a man in Albrecht who [in a way] is offering to save her when her husband, you could argue, left in the hour of need..." (PC)

As Sheers implies here, the moral issues associated with this occupation are not straightforward. Every decision to act involves both compromise and contradiction. For example, for the husbands to protect their country through enacting their auxiliary units, they had to leave their valley; however, if they had stayed it was likely that military action would have arrived with an inevitable loss of life. Similarly, for Sarah and Albrecht, the former is not a straightforward victim of circumstance and the latter is not a ruthless invader (with the concomitant simplistic values associated with such binary positions); rather the novel seeks to complicate such straightforward dualisms. Albrecht for example is leader of a patrol of 'broken souls' (Sheers, 2007:39); they have been fighting across Europe for the duration of the war, and no longer want to continue the bloodshed. Their arrival in the Olchon offers an opportunity for respite from the frontline; to them it seemed as if:

> "... they'd arrived in a rare heaven. After the years of field living, sleeping in dugouts, on wooden plank beds, in fox holes, a place like this was a Utopian dream. It made Albrecht feel clumsy, awkward, as if they were a pack of stray animals wandered into an empty palace. Their boots, their weapons, their talk, they all seemed too modern and harsh for this ancient [valley]." (Sheers, 2007:114)

The geography of the place – its age, culture, and beauty – played a key role in suggesting to Albrecht that for him and his patrol, the valley might offer a hiatus from the war, a place to recover their humanity and hope. Albrecht imagined that:

> "this valley could be different. He'd already begun to feel the faintest of turnings within himself this past week. He knew it was the valley

> that had engaged this turning and he wanted it to continue, this slow rotation inside him like the tumblers of a lock edging into place. If it went on for long enough, until the end of the war, then who knows? It might just unlock him altogether. It might still not be too late and he could finish his war a victor in every way. Over his enemies and over himself." (Sheers, 2007:137)

In this way, locating the story in a landscape Sheers knows 'down to the bone' enabled the author to make "the landscape ... one of the most important characters" in the novel (Sheers, in Griffiths, 2011). As a place of refuge, the land reminds Albrecht of his humanity, in Sheers' words, it "re-awakens him to the man he was before he was a soldier" (ibid.). Indeed, the physical remoteness of the valley was something chosen explicitly by Sheers to entangle with and affect the plot. Sheers had identified a particular history to the valley whose remoteness seemed to encourage a utopian atmosphere, as he outlined as we continued our walk:

> "because of the isolation of the valleys, groups of people [over history] have felt that they could live a different way here. Of course probably the first group of people to really pick up on that idea were people who built the Priory down there, a very isolated spot to live in a different way, and again legend has it that they were following St. David who had come here again [to find] a place of reflection, a place to find themselves again..." (PC)

Figure 11.3 The Olchon Valley: a credible and believable isolation. (Source: Author)

Playing with this idea of retreat and escape was possible for Sheers not simply due to the history of the valley, but also due to its ongoing isolation. The Olchon Valley remains poorly served in terms of vehicular and as well as telephone communication, and as a consequence those in the valley kept themselves very much to themselves; as Sheers outlines in his story:

> "They had no phone lines this far up and apart from Maggie's farm, which sat higher in the valley, the whole area was dead for radio reception." (Sheers, 2007:7)

> "This was a backward valley, a dead-end, isolated slice of life embedded in the clefts of the hills. Even the information [Albrecht had] gathered in the last village indicated these hill women were considered a breed apart, outsiders.
>
> 'Well, they'uh a quiet lot, tha's foshaw. 'Ide behin' th'edge if they see yer, they do'." (Sheers, 2007:110)

Walking along the top of Offas Dyke, Sheers explains how this remoteness was vital for how the plot progressed:

> "the whole premise rested on the fact that this isolation had to be credible and a believable isolation ... in a period when still an awful lot of travel was often done on foot or by horseback, you can understand why going from down there in Llanthony up to over there in the Olchon was no small feat. Even by road you're talking about going right of the valley and around the very end of the Hatterall Ridge [over there]: [you can] feel that isolation. I talked to as many people who knew this area as possible in the 1940s and that was very much confirmed, and specifically about the Olchon Valley ... it is kind of a forgotten valley. Very strange that isn't it? It's a proper border territory and I suppose as soon as I went there I felt that I could find that necessary isolation that the novel very much needs." (PC)

In many senses the Olchon was an isolated, remote, and 'liminal' space (see also Chapter 5); a valley forgotten by local villages and towns, and ignored by the outside world. This was a valley located precisely in 'border country',[2] falling between the folds in the hills between Wales and England. When Sarah moved to the Olchon to marry her partner Tom her parents called the location 'the last valley in Wales', whilst her father teased her that it was the 'first in England' (Sheers, 2007:64). With no country wanting to claim this space, the

---

2 As Sheers explains, the notion of this area being a world apart from England and Wales is a popular one: "Quite a few people round here will actually refer to themselves as borderists more than English and Welsh, I think it's a very interesting mind-set to investigate" (Sheers, 2011).

Olchon appeared to have fallen off the map. Indeed that was exactly the case for Albrecht's patrol.

> "Their position [in the valley] lay on the border of two command sectors. The commanders of each sector would be all too quick to shift responsibility for a nomadic patrol onto the other's shoulders. The army supply lines were stretched, they were fighting a desperate, defending nation. Winter had come upon them like a second attacking front. Any concerns they might have had about a patrol in their sector would have been drowned out in the noise and confusion of war. Albrecht and his men had inadvertently slipped off the stage of battle." (Sheers, 2007:188)

The Olchon valley, hidden in the folds in the mountains and the folds in the map of the German army, became somewhere where the characters could, like St David and others before them, try to find a different way to live out the war.

Box 11.2 The Olchon

Owen Sheers borrowed from the lore of the Olchon area to describe the geography of this isolated valley, as well as the particular relations to landscape generated in this border country. As he says:

> "I have been influenced by Raymond Williams and specifically his trilogy, 'People of the Black Mountain', in all sorts of ways. ...Actually I purposely used his description of the Black Mountains as a hand, and it's something that Bruce Chatwin uses in 'On the Black Hill' because he picked it up from Raymond Williams, but actually you know there's no surprise, it was a way of describing these Black Mountains that several people I've spoken to have heard their parents and their grandparents talk about as well. I loved that idea that there was this very ancient idea of the Black Mountains as being like a hand, so the Olchon would be between the thumb and forefinger, and I sort of wanted to continue that on as a nod to this existing literature." (PC)

In *Resistance* this local lore is introduced in the following way:

> "When Sarah was a girl her mother had once described this range of long hills as a giant outstretched hand. It wasn't an original description – Sarah had often heard people refer to the 'Black Mountains Hand' ...She'd had a deep fondness, a belief almost, in such figurative language. As she got older these phrases became her hand holds, a semiotic map by which she navigated her way through the days, weeks and seasons." (Sheers, 2007:11)

> "Half-way up the slope Sarah eased Bess to a halt and turned to look back down the long V of the valley. If she were to follow her mother's description of the Black Mountains Hand, then where she rode now was in the hollow between the thumb and forefinger. The Hatterall Ridge, on her right, was the forefinger, a long slice of land pointing south-east towards Pandy with Offa's Dyke running the length of its knuckles." (Sheers, 2007:63)

Using this lore to describe the remoteness of the landscape is not just a rhetorical device, but points toward the lived experience of isolation generated by being in this place. Indeed, due to the uniqueness of the valley the farmers had created a strong lifescape in the area. As we have seen in Chapter 9, rural lifescapes are formed by the "social, cultural and economic interactions between humans, livestock and landscapes... and in turn shape ways of being in the world" (after Convery, 2005:100). In *Resistance*, it was the men who predominantly worked the land and led the creation of these lifescapes, however in their absence the women had to make sure they were sustained; as Sarah experienced:

> "And they were good dogs too. .... never drove the flock too fast. Could bring them all the way off the mountain just by standing, poised, in the exact positions where Tom placed them. How could she ever hope to replicate that? With her the dogs associated the hill with rides ... or when she went picking bilberries, when they could loop out and back to her in a loose pattern of coming and going. With Tom the hill was their place of work, a tight map of pressure and release, of give and take as the flock flowed before them like a shoal, three hundred animals moving as one." (Sheers, 2007:67)

These lifescapes, along with the 'semiotic' maps used to describe the land, led to a particular way of thinking and way of being in the Olchon. As a writer who knew this place intimately, Sheers was sensitive to this and transposed this intimate coingredience to his characters; as we see in the following examples of Sarah, and firstly Edith, whose husband and sons had left for the resistance, but who would not leave her farm in the hills:

> "They'd made sure someone visited Edith every day. She was still fiercely independent and wouldn't hear of moving down into the valley. With her boys gone Maggie would have had room for her. And Mary too. But Edith wouldn't leave The Gaer. They were retreating into the hillside together, her and the cottage, sinking back into the soil." (Sheers, 2007:58)

> "The hill fort itself was now no more than a series of faint concentric rings buried beneath centuries of soil and grass. It was as subtle a feature on the ridge as the banks and dips of Tom's body had been in the horsehair mattress of Sarah's bed. Like Tom's outline the missing physical presence of the fort, its ramparts and defences, could be traced only by someone who knew the place intimately, who could still see what was no longer there in the earth echoes underfoot. A careful eye, sensitive to the landscape, could make out where a gate once stood or the foundations of huts where men had once slept and fought and loved and cooked. To the casual observer, however, there was nothing there, just a toothless gap in a long grassy jawbone of earth and a few faint humps beneath a tangled mass of bracken and gorse." (Sheers, 2007:56)

Power, isolation, and interdependence

As Foucault (1980) tells us, no system of power is ever absolute – it is always possible to resist, escape or subvert a system of domination. In this view, the Olchon came to represent a space in which the World War had come, the German state had sought to impose their ideologies on the British nation, but had failed to wholly dominate. Hiding between the folds between England and Wales, the Olchon had become somewhere which functioned as a geographical equivalent to the 'cracks and vacancies' in any system of power (Bey, 2011:99). In this case, the valley had become a 'temporary autonomous zone' outside the ambit of war. According to Hakim Bey, a Temporary Autonomous Zone (or TAZ) is a space which 'liberates an area' for a period of time; the zone is perhaps perfectly suited to a remote location such as the Olchon as it survives best where it:

> "... can 'occupy' areas clandestinely and carry on ... for quite a while in relative peace. Perhaps certain small TAZs have lasted whole lifetimes because they went unnoticed, like hillbilly enclaves – because they never intersected with the [State].... the TAZ['s] greatest strength lies in its invisibility.... TAZs [come to be] either through neglect on the part of the State or because they have somehow escaped notice by the mapmakers." (Bey, 2011:102)

Temporary Autonomous Zones are thus areas where alternative ways of living can be actively practiced due to the absence of wider systems of government or domination. In this space, the women of the valley and the patrol could choose how they wished to co-exist. Initially, they had decided to lead parallel lives, sharing the same space but choosing not to engage, overlap or meaningfully integrate (after Community Cohesion Review Team 2004);

> "Sarah resisted the assistance of the patrol the most. ... she'd been especially wary of Albrecht and wanted no more contact with him and his men than was absolutely necessary." (Sheers, 2007: 197)

When the patrol first arrived, the women and the soldiers effectively stood apart, "standing with their backs to each other" (Steiner, 2001: 145, cited in Kymlicka, 2003). However, as the winter snows came, the non-human geography of the location conspired to ensure this stand-off couldn't last. As Sheers writes:

> "No road had ever run through the Olchon and now it was as if no road had ever run into it either. Absorbed in their work, in thoughts of their husbands, the women had become amputated from the world beyond. It was as if it had forgotten they ever existed. ... the snow had perfected their isolation." (Sheers, 2007:164)

Using the geography of the location, Sheers forces his characters "into a situation of mutual dependency" (PC). The presence of snow, alongside the necessities of maintaining the lifescape, meant the women had to call upon the soldiers for help:

> "... the farm demanded [Sarah] as a suckling baby demands its mother; without malice, without agenda, just simply because it was the way of things. ... The whole valley was there, with its cycle of birth, sowing, harvesting and slaughter, and they, the women, had to keep it turning or it would leave them behind altogether." (Sheers, 2007:77)

> "[So] here they [the soldiers] were again. In her house, in the kitchen. The enemy. The invading army. But it was different this time. As if the snow had shed them all of their history. She felt safe. For the first time in over a month, she felt safe." (Sheers, 2007:155)

The advent of the snow thus changed the situation for the valley. Initially the relations between the women and patrol were couched in terms of occupation and (r)evolution, they were enemies 'red in tooth and claw' (after Tennyson, 1850). However, winter brought a new scenario; now in order to survive their relations necessitated 'mutual aid' (after Kropotkin, 1915). The transition to this enforced interdependence was not easy. For example, since Tom had left, Sarah had maintained a (one-way) conversation with him through noting thoughts and events in a ledger; this change in relations with the soldiers she found hard to articulate, like many concessions to the enemy in the wider world, it felt like "the most personal of betrayals" (Sheers, 2007:218):

> "Sarah laid the book and the pen on the table and went to stir the *cawl* simmering on the range. They hadn't been able to kill the sow on their own, she was too strong. It was Alex who had got her onto the bench. And Alex who had strung her up afterwards. But Sarah couldn't write that to Tom. She had no way of telling him without it sounding wrong. So she left it out instead, as if it had never happened." (Sheers, 2008:221)

Over the winter, this mutual aid and interdependence became part of the lifescape of the valley:

> "At first Albrecht had been careful to limit the contact as much as he could, but now the dependence between the women and the patrol had developed its own rhythm and metre. He was no more the conductor of events, but merely an observer .... Slowly, despite the language barriers, he felt the patrol were becoming less German in the women's eyes and more just men. Men who'd been washed up on their doorsteps, carrying with them their own losses, just as the women did the loss of their husbands.
>
> This shadow of their missing husbands was, of course, always there, and made darker by the intermittent wireless reports of the ongoing insurgency. But even this was now woven into the everyday fabric; an unseen and mostly unspoken weight of absence which together with the war worked as a counter-balance to their lives in the valley." (Sheers, 2007:272)

The domination of the German patrol and the resistance of the Olchon women is not oppositional or dualistic here. Although the women are deserted, they are not victims; they exploit the patrol to enable their farms to function. Despite the patrol initially being dominant colonisers, in practice the valley gives them back their humanity and mends their souls, as they keep the cycle of the lifescape turning. The power of each group here is therefore not reduced to a simplistic 'good' or 'bad' (related to singular judgements of *Resistance* or 'domination'), but Sheers makes it clear that power here is entangled – both groups act in ways which can be seen to simultaneously dominate and resist the other, as they both suffer from the onset of war and have to find an appropriate way to negotiate their situation. Each group navigates the tension between staying loyal to their orders, as well as their roots; each attempts to create or protect the(ir) world, "or at least an idea of it" (Sheers, 2007:234).

The world and how it should be

These key questions are brought into focus through a plot device that reveals the patrol's real reason for their presence in the Olchon: the search for the Mappa Mundi, a rare medieval map of the world. In *Resistance*, the Mappa Mundi was a much prized objective of the Third Reich; if the Germans owned the maps (so the argument went), they would own the world. In the factual version of history (rather than the counterfactual one), "the Mappa Mundi was removed from Hereford Cathedral for the duration of the Second World War and, after a stint in the cellars of Hampton Court, kept in a coal mine in Bradford-on-Avon" (Sheers, 2007:354). For Sheers' counterfactual history, the author felt the reader "could just about believe that it might be shunted north, and almost brought back to its home and hidden somewhere [in the mine tunnels] around the Olchon" (PC). For Sheers, this map emphasises the vital role of place, roots, and identity for the characters; as we sit on the Llanthony side of Offas Dyke in the late afternoon sunshine, he explains that:

> "the most important lines in the book are when Sarah asks Albrecht 'what is this [the Mappa Mundi]?' in that question is a struggle they continue to have about the idea of the world and how it should be. To what extent do you change with the world , to what extent do you

resist and say 'I'm not actually going to change', so the map actually went on to become a very important figurative engine." (PC)

Ultimately, and perhaps importantly through a lifescape metaphor, Sheers shows that certain lines cannot be crossed by Sarah and Albrecht. Although over the winter they had become mutually dependent, their roots and futures were always going to be holding different lines. This is explained through Sarah and Albrecht attempting to convince a ewe to mother an orphaned lamb through covering it in another sheep-skin:

> "The lamb reached the ewe and began butting its head in the crook of her hind leg. The ewe, in response, bent to smell the lamb's back. She lingered over the skin, scenting the dried remains of her own amniotic fluid. The lamb, shaky on its legs, kept up its jutting motion under her stomach. The ewe sniffed once more about its tail, then lifted her head and walked away to graze again.
>
> Sarah let out a heavy sigh. 'She's rejected it,' she said, shaking her head. 'Even with the bloody skin on, she didn't take it.'
>
> The lamb, bleating thinly and shaking its tail followed the ewe, but when it reached her she walked away again, oblivious to its cries as if she were alone in the pen." (Sheers, 2007:258)

Albrecht, the sheep in wolf's clothing, as well as the German in Tom's jacket, was ultimately rejected by Sarah. Their temporary autonomous zone was just that, temporary, and as the snows melted the war came back to the Olchon.

> "'What are you doing here?' she said. Her voice was clean and hard as bone, pared of all depth by her anger. Her eyes remained locked on his. 'You've won this war. Why don't you just leave now? Leave us alone.'
>
> It was no good, Albrecht saw that now. He was not, as he had hoped, the sum of his parts, of the parts of himself he'd rediscovered in this valley. He was, after all, no more than the sum of this uniform, of this war. He had no power to shape his future. He could not extract himself, escape from what he had done and what he had been part of, through no will of his own, over these past five years." (Sheers, 2007:230)

Conclusion

In the counterfactual history of *Resistance*, Sheers has entangled place and imagination in order to explore the implications of war for identity and belonging. As Sheers states:

> "The epigraph of the story is to do with the concept of Hiraeth and the sense of longing for something or somewhere you may have never owned. ...My interest in landscape in the novel was in my engagement with places and the emotions evoked when I'm in the Black Mountains. I think the importance is that the landscape isn't just where the Welsh live and inhabit – it becomes a place of refuge but also of danger. On the other side – the German soldiers – it's the landscape which unlocks the main character, Albrecht, and re-awakens him to the man he was before he was a soldier. And the people in the valley are kind of forced to find a position of mutual dependency which raises up a question mark over the concept of war. It is an old question, but one which still raises the idea that the enemy has as much humanity as we do." (Sheers, in Griffiths, 2011:no page)

In *Resistance*, Sheers explores ideas of "sacrifice, collaboration and occupation" (Sheers, 2011: no page), examining the "complex territory of where the lines between resistance and collaboration blur" (Sheers, in Jones, 2011: no page). More than this, however, *Resistance* focuses on the vital relationship the characters have to the land, in Sheers' words, "the story grows out of the landscape and the relationship to the land" (Sheers, in Griffiths, 2011: no page). In the same way as in his poetry, this relationship to the land gives meaning, identity, and belonging to people, it gives a sense of humanity, and of a place in the lifescape. Through the "conversation / of place and page over time", Sheers shows how the relationship to land can give insight and identity in order that "each / in turn, might hold the line". We see how for Albrecht the land was capable of reminding him of the man he was before the sacrifices of war, whilst Sarah looks to the lifescape for reassurance and roots in the absence of Tom and peace. As a consequence of these connections we can see that where rooted relationships to land *places individuals*, war inevitably *displaces individuals,* be they either the invaders or the colonised. As illustrated by both Sarah and Albrect at the end of the novel, war has the effect of undermining co-ingredient relations, cutting the ties between people and place so in the end, there is no victory for any side.

"'I am German,' he said, turning his face away from the sun and looking back down at the abandoned priory. 'But Germany is not my home any more.'

For some minutes they sat there, beneath the tree, saying nothing. A blackbird sang from one of the ruined walls beside them. A crow cawed, brash and harsh, as it fought in the tree above with a pair of magpies. It was going to be a warm day. Sarah could feel it, the potential heat heavy in the air. She was in the valley of her childhood, where she'd grown up. She knew the contours of the land about them intimately. And yet she was also in a place unfamiliar to her, somewhere off the edges of any map she'd ever known." (Sheers, 2007:283)

Chapter 12
Geography is destiny: Who and where is Iain Sinclair?

I get off the train in Hackney, East London, on a balmy spring day. I am en route to meet Iain Sinclair at his home in order to discuss his novel *Landor's Tower* (2001). Nestled in my pocket is a copy of *Ghost Milk*, Sinclair's chronicle of corporate enclosure in his adopted Hackney neighbourhood (2012). With Sinclair's arguments fresh in my mind I walk from the Tube and try to gain a sense of the changes occurring in this borough and how they may affect the lives of its inhabitants. As we will see, Sinclair has become famous as a chronicler of change in London, but perhaps less well known is his origins in and fiction about Wales. As we will see, however, these origins may be responsible for his current endeavours.

> "I have no clear memory of the event, even though I was already two or three days old when it happened. What I do remember is my mother telling her version of the tale. Repeated often enough, these fables become history. Out of an enveloping plain of darkness that absorbed all mere facts, she highlighted a few over-directed and non-linear tableaux.
>
> The ambulance hesitates ... on the rim of a hill overlooking the city of Cardiff. (Premature visions of future horror: driving tests, relatives in sudden and unexplained hospital deaths, poetry readings.) The pregnancy was a tricky one ... mother and new soul, suffered a long, painful parenthesis – and [then] I was born: out, present. Roath Park, or thereabouts. (Memories of this lake, water and vegetation, will be transported to London. A lifelong fondness for the spaces of Victoria Park, Hackney.) Cardiff was not where we lived, I was estranged from it, to a degree that might seem overdone: I refused to take a breath. Despite the warmth of the slaps and blows that induct us into the social world, I was stubborn, blue as lead. As a Bluebird (the slow crucifixion of Ninian Park fandom).
>
> It didn't look good.
>
> My father paid a visit, a small pile of paperbacks crooked under his arm. A typically oblique but significant gesture. Entering the heat and nervous closeness of that private cell, he stumbled. A tall man, stiff Highland bones, he dipped dutifully to look into the cot. A single volume dislodged itself, fell through the air and struck the baby on its undefended (still open) skull. Plucked from deep dreams the newborn creature howled, sucked recycled breath into reluctant lungs.

> Launched, there was no remission.
>
> Cardiff-born: for life.
>
> On every form, passport, criminal record.
>
> My mother held on to that title, where she confused all others: *Crackerjack...*" (Sinclair, 2005:189)

In this remnant from his life- and short-story, the themes of Sinclair's more famous work echo strongly: the importance of connections to geographical place, the associated and perhaps inevitable estrangement from these roots, the entanglement of fiction into reality (and how in this case fiction can even bring life to person-place relations), and the preference for 'non-linear' narrative structure. From these Cardiff roots, Sinclair has become identified as a 'city stalker' and a 'culture hustler' (Sinclair, 2012b:13). Using the premises of psychogeography (see Chapter 1 and below) but reinterpreting them in his own style, Sinclair has built a reputation for exploiting his "gift for bullshit, [and] never letting facts get in the way of a good story" (Sinclair, 2012b:17). From Chobham Farm in London in the 1960s, he has focused his interest in chronicling places that have been overlooked or underwritten; Sinclair's places are passing places – places that have been bypassed, are passing, or have (been) passed over. Changing *no-wheres* to *now-heres*, Sinclair's role has been to put a value on these places in these states. In his words, these passing places "seduced me, [they were] unwritten nowhere[s] in which to launch a lifelong journal" (Sinclair, 2012b:25).

Urban pathology

With this in mind, the critic Rachel Cooke (2009:no page) suggests Sinclair is a "writerly pathologist, always in search of decay, the better to note the time of death". The moniker of urban pathologist suits him well; he is the-not-so-silent witness to the passing of the geographies of the unnoticed and unloved. Sinclair gives these geographies a few last beats of life, turning our attention to them at the point where we can do no more for them. The theme of death seems to echo Sinclair's oeuvre: according to Adams, Sinclair's writing is "an act of vigorous exhumation; he fleshes out extraordinary contemporary satires from an unlikely rubble of old bones" (2001:no page); and in the words of

Penman, "Iain Sinclair writes ghost stories, of a sort: whatever his subject, there is always a low, persistent note of something mourned, spectral, lost" (2001: no page). As the following excerpt from *Ghost Milk* explains, embalming the lost is Sinclair's metier:

> "'You have a name for your book?' Mimi said. '*Ghost Milk.* What does this mean?'
>
> 'CGI smears on the blue fence. Real juice from a virtual host. Embalming fluid. A soup of photographic negatives. Soul food for the dead. The universal element in which we sink and swim.'
>
> 'Crazy, Mr Sinclair,' Mimi said. 'Crazy again'." (Sinclair, 2012b: 338)

Perhaps Mimi would agree with Sinclair when he describes himself not as a psychogeographer but as a "psychotic geographer" (see Coverley, 2006); nevertheless in line with the psychogeographical approach there is an element of nostalgia, wistfulness, and politics in his writing. As we have seen in Chapter 1, it is argued that much psychogeography has a "radical nostalgia" to it, it is at once "funny yet melancholic, utterly of our time but ill at ease with modern Britain" (Bonnett, 2009:46). Sinclair's psychogeography is thus anti-establishment in its politics, particularly if that establishment is one defined by a post-Thatcher ethos of rootless, mobile capital which imposes the "throwaway impermanence of the present" onto our lives (Sinclair, 2012b: Blurb). The anger in Sinclair's psychogeography is so near the surface that it can sometimes blind the reader to the more subtle connections to place in his work.

For Sinclair is a (psycho)geographer; and he practices this geography through the mechanisms used predominantly in this book: walking. As Sinclair puts it: "walking is the best way to explore and exploit the city... Drifting purposefully is the recommended mode" (cited in Pinder, 2005:396). He states that "urban walking is a way of contacting the ghosts and levels of a city, the past and the future" (cited in Pinder, 2001:12). Although his work does not share the complete political manifesto of the Situationists who initiated the practice (see Coverley, 2006; Pinder, 2005), he nevertheless seeks to politicise place through his "transgressive wanderings" and "re-imagine the forgotten nooks and crannies of ordinary landscapes to re-enchant and re-mythologize prosaic geographies" (Bonnett, 2009:46). Sinclair's walking is not the meandering of an arcade flâneur (see

Benjamin, 2002; de Certeau, 1984), but is more *à la mode* of a post-modern city stalker; as he writes:

> "The concept of 'strolling', aimless urban wandering, the flâneur, had been superseded. We had moved into the age of the stalker; journeys made with intent — sharp-eyed and unsponsored. The stalker was our role-model: purposed hiking, not dawdling, nor browsing. No time for the savouring of reflections in shop windows, admiration for the Art Nouveau ironwork, attractive matchboxes rescued from the gutter. This was walking with a thesis. With a prey... The stalker is a stroller who sweats, a stroller who knows where he is going, but not why or how." (Sinclair, 1997:75)

Sinclair uses city stalking as a way to create political art-walks/works, "creating a walk that would allow you to enter into a fiction" (cited in Coverley, 2006: 120). Resonantly echoing with the concept of plotlines, Sinclair's writing explicitly weaves geography into (moving) fictions, and as part of their entanglement, threads through personal history (as in the case of *Crackerjack*) as well as (re)interpreting literary figures and oral histories in order to lay a new weave onto a place (as we will see in *Landor's Tower* below). As Coverley states: topograph[ies are] reconstituted [by Sinclair] through a superimposition of local and literary history, autobiographical elements and poetic preoccupations, to create an idiosyncratic and highly personal vision" (2006:120).

Non linear tableaux

> "Exit Ghost: 'Walking the streets like a revenant, he makes connections that explode his carefully protected solitude'." (Sinclair, 2012b:398)

Sinclair's entangled geographies therefore link together the real and imagined, literature and place, alongside personal and collective histories. The manner in which these geographies are entangled are not orderly and systematic, but rather random and chaotic. As Penman accurately puts it:

> "Sinclair texts were like internet sites before the fact, full of links for the interested reader to follow. ...he built up an alt.rolodex of contemporary mystics and mummers." (2001:no page)

Like a human world wide web, Sinclair makes inspired and idiosyncratic connections through his wanderings, filling his pages with empirical documentation, political polemic and imaginative musings. In the same vein as the poet John Clare (who claimed that he didn't write his lines, he simply "found them in the fields and wrote them down" (1908)), there appears to be little sifting or filtering in his work, nothing akin to a conventional editorial process. His critic and sometime character Michael Moorcock (see below) writes how Sinclair's prose:

> "reminds me a little of the later, wilder imaginary conversations of John Cowper Powys, that Welsh genius who tried to put everything he had discovered and thought about the world into his books, cramming them to bursting with the wealth of his magnificent curiosity. Sinclair adds an element of experienced common sense, of amused relish, which in the end both grounds and humanises his impressive visionary flights." (Moorcock, 2001: no page)

In this way, Sinclair adopts an open-form structure for his own brand of psychogeographical mythology (see Sinclair, 2012b). Like the film 'Bronco Bullfrog' on which Sinclair played a minor role in the 1960s, his writings tend to be "unconstricted by a bullshit narrative arc or conventional three-act structure... Characters, remembering not acting, move around just enough to quantify a portrait of place" (Sinclair, 2012b:26). Sinclair writes how such a lack of structure is a "final renunciation of the burden of narrative. The fiddle of editing. What passed through the camera [or indeed the author's eye], that was the story" (Sinclair, 2012b:31). As a consequence, Sinclair's writing defies convention, and to an extent classification, through having a "blank page of an elephant folio" (ibid. :392) onto which he captures places, he had "blunder[ed] into a form of writing that was neither fiction nor non-fiction" (ibid.:88). As illustrated in *Landor's Tower*, when Sinclair (in one of his many guises) is asked whether he is writing a book or a film, the questioner realises that, with Sinclair, these categories "didn't matter, those nice distinctions no longer had any currency. Truth, fiction. Sports report, road kill statistic. [They] are all part of [his] spectacle" (Sinclair, 2001:25). The reader is left with an evocation of entangled relations between person-page-and-place through which telling and retelling is required in order to take it in and "metabolize it all" (ibid.:393).

Fear and Loathing in South Wales

In *Landor's Tower*, Sinclair employs his own brand of psychogeography[1] to produce a spectacle that most closely resembles a 'fictional' novel. *Landor's Tower* is in many senses a homecoming narrative. It begins when, "a writer, who has lived for years in London, reluctantly acknowledges his growing obsession with the Ewyas Valley" (Sinclair, 2001:blurb) (see Map12.1). The novel is thus Sinclair's chronicle of his own attempts to deal with his estrangement with South Wales, the land of his childhood.

As we have seen in Chapter 11, the Ewyas Valley and its surrounding area has been a site of "persistent attempts to found or imagine utopian communities, all fascinated with the mythology of the west..." (from Ibid.).[2] I ask Sinclair about the importance of mixing the idea of 'homecoming' with the notion of the 'frontier myth' in *Landor's Tower*:

> "In *Landor's Tower* my mindset was very much 'West focused', stepping westward you know, in a frontier sense, but it was also a kind of homecoming for me in lots of ways.... There's an element in Welsh writing that is oriented around the voyage, the interior and exterior voyage, the self-defining journey. It's been influential on me as I grew up in Wales and even when I moved to London ... it's been a big influence...defining yourself in a landscape you don't really know." (PC)

---

1 As Sinclair comments on psychogeography, "For me, it's a way of psychoanalysing the psychosis of the place in which I happen to live. I'm just exploiting it because I think it's a canny way to write about London. Now it's become the name of a column by Will Self, in which he seems to walk about the South Downs with a pipe, which has got absolutely nothing to do with psycho-geography. There's this awful sense that you've created a monster. In a way I've allowed myself to become this London brand. I've become a hack on my own mythology, which fascinates me. From there on in you can either go with it or subvert it" (in Coverley, 2006:122).

2 Walter Savage Landor's Tower itself, or the remains of it, can be found close to Llanthony Priory and Offa's Dyke (see Chapter 11).

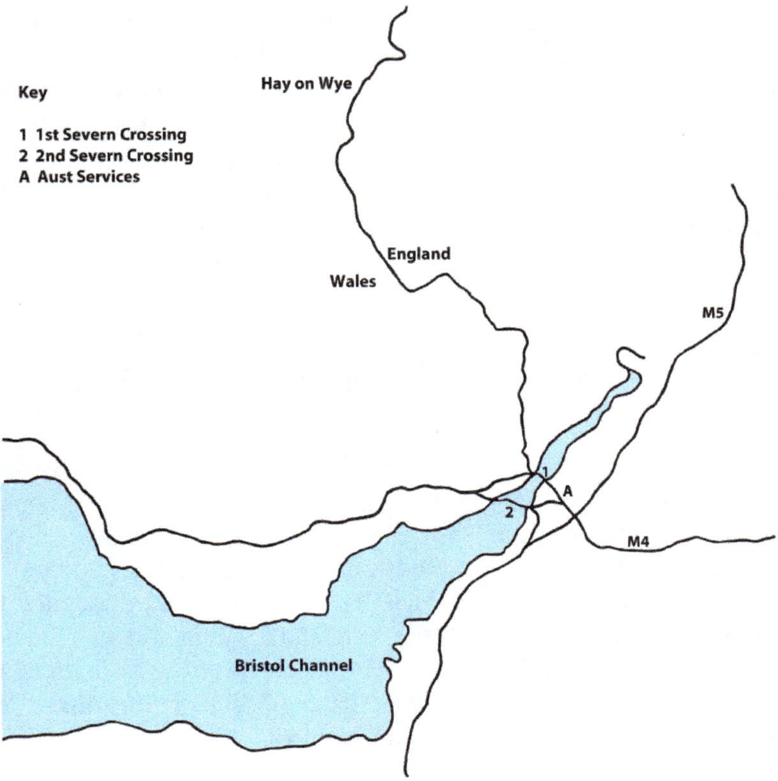

Map 12.1 Borderlands. The border of England and Wales, and the location of the Ewyas Valley.

In heading west towards the frontier of Wales, Sinclair is exploring the frontier of his youth and the (un)known geographies that define him. In returning to the landscape he grew up in, and perhaps grew out of, Sinclair revisits, reframes and reconstructs his person-place relations. Due to the constitutive co-ingredience of people and place (see Chapter 1), this process means Sinclair encounters past and present versions of himself that are brought forth by this new landscape. As we have seen in Chapter 5, humans do not have a monolithic, one dimensional identity, rather they have an identity that is fragmented, multiple, and often inconsistent. This identity is influenced by our location, and is liable to change as we move from place to place. In different places we therefore become different people as various aspects of our personality comes to the fore. As a consequence, identity is never a singular and unified thing, but rather is a bundle of multi-phrenic selves, influenced and formed through our geographical locations (see Featherstone, 1995; Anderson, 2004b). Sinclair takes this insight seriously, accepting that his identity is strategic and increasingly fractured, multiply constructed across intersecting, and often antagonistic, discourses and practices (after Hall, 1996, p. 4).

Box 12. 1 Hay on Wye: "I don't need a map. I'll track us to Hay. You can smell the fear twenty miles off" (Sinclair, 2002:6)

One of the key locations in Sinclair's border landscape is the town of Hay on Wye, the "extraordinary, half-mythical city of books" (Moorcock, 2001:no page). Perhaps unfairly, Hay is often described as the place 'where books go to die', principally due to the vast number of second hand bookshops which have become a key magnet for visitors and tourists. As a 'little England' within Wales, Sinclair conducts "the topographical equivalent of character assassination" on the town (Penman, 2001: no page). Adams (2001) suggests that, "few writers ...could summon such righteous savagery for [its] booksellers", as Sinclair (in the guise of Norton) writes:

> "In the past, I'd arrive at Hay... before the 9 a.m. opening time. I was hot to plunge in among the stacks of the Cinema Bookshop. Enthusiasm wilted with every yard of alphabetically arranged fodder; only by exhibiting so many books that nobody wanted would you disguise the magical sleepers. Or so the scouts told themselves, for the first couple of hours, as they hacked their way through shelf after shelf of titles that never make it out of used book pits... I'd drudge through the town, shop to shop, finding it harder as the years went by to get my hands on anything that proved my journey was necessary. A few gaudy paperback shockers, makeweights in container loads of US remainders, dug out of a cellar: junkies, axe murderers, teenage hoods, bikers, beatniks, tame outlaws. The occasional fantasy classic that slipped the net. Some deservedly forgotten Whitechapel hack waiting to be reinvented, reassessed, talked up." (2001:32-33)

Figure 12.1 Where 'books go to die': Hay on Wye. (Source: Author)

"The streets of the town were deserted, the tourists had not yet arrived... Narrow streets, a market perch for cheeses, charity caves and a clock tower. A small, steep hill running up to the ruin of a castle, which was surrounded by barrows of dead books, future kindling. Hay was bereaved, mourning past glories, resigned to its present shame.

'The town,' said the poet Anne Stevenson, 'is full of people who have limped this far and don't have the strength to limp any further'." (2001:37)

As Sinclair's narrator Norton suggests, "This was not the place in which to start work on a book. Who needs another trumpeting volume to chuck on the pyre, another dreary effusion to bulk out the melancholy mound? Hay was aversion therapy for bibliophiles." (2001:37)

"The nice thing about Hay, it was a good town to leave behind." (2001:60)

For Sinclair, the borderland between England and Wales is a landscape he should feel at home in through birthright and upbringing, yet he feels unearthed and uprooted from it through his migration to London. In his absence, the landscape has changed, and so has he. The novel seeks to plot Sinclair's attempts to locate and "acclimatise myself [in] to the quiddity of this pluralist landscape" (Sinclair, 2001:62). I ask him directly whether the book is intended to explore a kind of schizophrenic conversation between his London version and Welsh version of himself:

> "Very true, that's exactly it, all the schizophrenic aspects on the doublings of Welsh English personality. The joy to me comes through place. Place is the most important aspect in this process [of identity and plot (re)formation], this geography is my roots and my destiny." (PC)

Sinclair adopts a novel mechanism through which to explore this geography, he employs his own schizophrenic nature to cast characters which have their origins in aspects of his own personality. As he outlines:

> "I wanted to split aspects [of my personality up] and share them out between different characters and let them argue a single entity into being. Splitting the single entity [of self] up into other things and hopefully reconfiguring it and forming it together to a new mythology and even a cosmology at the end." (PC)

Landor's Tower is thus a novel which is inhabited by many characters, several of whom are "essentially raddled and broken alter-egos [of Iain Sinclair]... [who] all lead you inexorably back to the blotted copybooks of their creator" (Adams, 2001: no page).[3] In creating such 'character-authors' (after Watts, in Stern, 2009:XII), Sinclair is performing "a sort of ventriloquism" (Sinclair, 2001:8). As Norton, the narrator of *Landor's Tower* states, "I could have [the character] Silverfish say anything I fancied, but I would still be talking to myself" (Sinclair, 2001:8).[4]

---

3 With this in mind, it is no coincidence that the subtitle to the novel is 'The Imaginary Conversations'.

4 Although it is true to say that Sinclair has created character-authors from his own identity, as we will see later in this chapter, the extent to which Sinclair retains any control over these characters is up for debate.

The location of *Landor's Tower* in the borderlands between England and Wales is crucial for this exploration of the schizophrenic relations between person and place. As we have seen in Chapter 11, this area is unique in terms of its ability to inspire utopian communities, but it is also significant because it can be a relatively isolated place. Coupled to these characteristics is the ease with which this borderland can be reached, and crossed, by the traveller, as Sinclair explains:

> "The whole exciting notion of borderlands [here] is the permeability, passing through the borderline that to-ing and frowing between Wales and England that people can reinvent themselves and reinvent their identities, they can escape and transgress." (PC)

As we will see, the geography of the novel is thus a vital figurative engine for Sinclair. It is as if the borderland offers a space outside conventional logics where it is possible for the various aspects that make up his identity to break apart and encounter each other. Beyond this facility however, the borderland space enables Sinclair to subvert and bring together a range of other convention-defying logics into his plotline. As we will discuss in detail below, the borderland space allows Sinclair's plot to 'dissolve the line' between the past and present and the real and imaginary. As Adams sums up, "Sinclair's fictional method is to hang a discrete series of improvised intellectual riffs [on these different elements of the borderland space]" (2001: no page). This has produced in *Landor's Tower*, a "memory movie, a sequence of prompts and quotations unpicking what we think we know" about the geography and identity of the author (Sinclair, 2012b: no page).

"Fucking writer! You've blown it. No plot. You can't do a road movie on foot" (Sinclair, 2001: 257)

Despite being a psychogeographer, Sinclair's journey in *Landor's Tower* is not a walking story. As his character-author Silverfish states, "You can't do a road movie on foot". In *Landor's Tower* walking is part of a palette of mobility that creates Sinclair's composition of place. His character-authors Silverfish and Dryfield (who we will see below have featured in previous Sinclair books), move through the landscape in the cocoon of a car. Through these aspects of himself Sinclair riles and goads not only his critics but also the psychogeographical aspects of his identity:

> "In the car, a heap of books on the passenger seat, he talked to himself: ricocheting fragments of reality, mottoes, on hoardings; the brain abdicating all sense of location.
>
> Silverfish tore the upside-down map from Dryfeld's grip…
>
> 'Brilliant. I entrust the map to a guy who's colour blind. That blue thing is a river, you asshole, not a road. That's the fucking Wye.'
>
> …
>
> 'You want to be a map fetishist, help yourself. I'm gone'." (Sinclair, 2001:5)

Here Sinclair satirises those who wholly rely on a dry, factual, geography in order to locate and direct themselves (he does not want to simply be a 'map fetishist'). In contrast, Sinclair prefers the interpretations of imagination and reality that are "more interesting than the near factual statistical documentation that doesn't really give you anything…" (PC). However he is also critiquing those who detach themselves completely from the 'real' through their literal and figurative automotive mobility. I ask Sinclair whether it was intentional to disconnect Silverfish and Dryfield from the world by locating them in the 'bubble' of a car:

> "Yes, they are tunnel vision obsessives, they have this vision of the world which is interpretable through fetishised totems – books or items of exchange or money. They could have violent philosophical arguments and debates with each other over cups of coffee or anything but they do not step outside the pod; they do not see where they are, they have no idea what's outside the window apart from the railway station, café, bookshop, but that's it. I see that as a form of madness, a sort of parody of the Thatcherite economics reduced to its absurd level for people who are behaving like that, disappeared from the world." (PC)

As Sinclair puts it in *Landor's Tower*, for these 'tunnel vision obsessives' the world is a place they are sealed away from:

> "Cardiff was asleep [as they passed through it in the car]. Silent avenues, private hospitals, old folks' homes: smeared, rubbed out by a single windscreen wiper." (Sinclair, 2001:328)

> "They ate (slept when they could) on the move. The car was ankledeep in burger cartons and jerk chicken wrappers, insulin syrettes, Moose Head lager caps (Silverfish); silver foil dishes with bean curds and threads of sprouting stuff (Dryfeld). It was an ashtray, a depository for ravaged newsprint." (Sinclair, 2001:6)

> "Dryfield loathed the countryside; between bookshops he would bury his head in catalogues. His sense of geography was functional and perverse: Chinese take-aways (the only places where he watched television), newsagents, grease caffs, charity shops (for ties, jackets, shoes), barbers (Trumper's of Jermyn Street), swimming pools, railway stations, massage parlours, morticians. The rest was invisible." (Sinclair, 2001:11)

Other aspects of Sinclair's identity were expressed through characters such as Norton and Vaughn. Rather than being confirmed car dwellers, these characters were closer egos to Sinclair's own altar of worship, the gods and mystics of psychogeography. In these characters Sinclair satirises his own obsession with walking and place, as well as reflecting on the capriciousness of success in the writing field:

> "Vaughan walked compulsively, roamed the local hills, scribbling precise meteorological observations, reams of stuff about light shifts, psychogeographical displacements. His large head, hair aflame, bobbed above dry-stone walls. He talked to himself, yelping with excitement; the rush that signalled the next dip into catastrophic, knees-to-chin melancholia." (Sinclair, 2001:218)

> "I [in the guise of Norton] bored critics who once patronised me as a rough-trade novelty act, a dirty walker. I had the potential, so they said, to become a WH. Davies super-tramp with a diploma in psychogeography; now I was damned for sticking too long to the same midden, riverine London. The next stage, I knew all too well, was oblivion. I'd be lucky to have my name misspelt in a Time Out round-up." (Sinclair, 2001:26)

Here we see Sinclair not only as postmodern writer but also as critic, intensely self aware of his own multiple positioning as well as the tendency of literary media to pigeon hole authors despite these categories being inherently leaky and often redundant. Perhaps because of his penchant for the post-modern, Sinclair continues to play with these categories, and his ability to transcend them, through both self-critique and parody:

> "I was a trudger, steady, grim, lurching to favour whichever knee or, calf or Achilles tendon was currently out of commission. Moderate-pain was as comfortable as it was going to get." (Sinclair, 2001:62)

> "The crunch with this book ... came when ... I was reluctantly beginning to identify... with the novelist anti-hero (dissatisfied grafter, nice wife, beautiful daughters, working out his traumas by exploiting everybody he'd ever met in the name of fiction)." (Sinclair, 2001:37)

Paradox and Topography

> "Throw in rain that wasn't rain, saturated air, and you have a proper sense of a walk in the Black Mountains, jets screamed out of the half-mist, aborting lambs, making windows rattle in dark cottages. This, they reckoned, was as close at it comes to the Falklands." (Sinclair, 2001:62)

In the process of walking and driving the reader through this land, Sinclair locates himself in the mi(d)st of borderland paradoxes. Although the land is in one sense empty of culture and narrative (as he writes: "The solitude. The expanse of underexploited, unwritten land" (Sinclair, 2001:135)), it is also the location of writings and art by Wordsworth and Turner and, as a consequence, suffers from "mythological overload" (Sinclair, 2001:37[5]). In this paradox of cultural absence and presence, Sinclair composes his own plot by simply "pick[ing these cultural traces] up in any order and invent[ing] a narrative" (Sinclair, 2001:158). However, as Sinclair maintained through interview, the place remained constant in this selection and recomposition; his notion was:

> "that space is a sort of given ...You set out a space of geography you can operate in. You can raid it['s history], and respond to it, but [if you attempt to] change it somehow the place itself resists." (PC)

Geography for Sinclair is thus an active and limiting context. Place is part of the action and to a certain extent a co-author in composing the literary plot. However, in terms of the narrative content occurring in this space, the order of geography blurs into disordered genre and

5 As Sinclair puts it, "Doing anything fictional with Llanthony would be tautologous" (Sinclair, 2001:144).

convention; as Norton realises midway through the narrative: "I was tracing a palimpsest of David Jones, the Mabinogion and classic Welsh mythology, and found myself, up to the oxters, in Mickey Spillane" (Sinclair, 2001:287). In *Landor's Tower* time is similarly spliced and diced by the author, as Sinclair puts it, unlike space which 'is a given',

> "time is liquid and you can move freely through [it]. ... If you get to know that geography then whether it's happening in 1500 or 2010 makes no odds, it's all a continual present and you just move in and out of it. All of those [historical] elements are active in actually a piece of ground, which obviously itself is going through all kinds of [geological] processes. That's still the generator of imagination, the ground itself." (PC)

Here Sinclair suggests that time in a borderland space is a permeable medium. It is as if the palimpsests of plot that have occurred in the borderland are liquid layers into which one can plunge. In other words, in this space time is depth which can be accessed by the imaginative scuba diver. If this is possible, then place is actively reproduced as these different histories, and the people associated with them, are brought into co-existence through the plot. In *Landor's Tower* such a process is realised as Sinclair connects together different fictional characters, as well as historical or contemporary real people, who are swimming in this geographical context (see also below). As he writes:

> "Walking the ridge at the head of the Ewyas Valley, I felt the presence of other walkers; Eric Gill's impatience, the exuberance of the young Victorian curate Francis Kilvert." (Sinclair, 2001:64)

> "Arthur's Quoit. Which Arthur? Arthur Llewelyn Jones who became Arthur Machen? Tennyson's King Arthur? The bearded, laureate was here, he stayed at Caerleon. They all did, they passed through, leaving their tracks, snail trails. They must be acknowledged. My great-grandfather, it was his Christian name: Arthur Norton, Fellow of the Royal Colonial Society, Member of the Aberdeen Philosophical Society. Etc., Etc." (Sinclair, 2001:14)

In this conceptualisation, borderland space is an ongoing composition of tracks and snail trails, refreshed by the wakes generated by liquid history. Due to this framing, Sinclair is able to splice together stories which create mythologies for this particular place, and the characters that co-constitute it, as he states in conversation:

> "every single book [or plot] written in the past acquires a cast of characters who you are responsible for; so you've got to continue to listen to what they do as they evolve away from you or come back. For example the Dryfeld character we talked about here first appeared to me in *White Chapell Scarlet Tracings,* one of a number of crazed booksellers who are very specific to the London markets of that era.... Then as time goes on you get to know the character a bit better and discover he is even stranger than he first appeared and he's now got a car and he's moving about in the Welsh borders; it becomes critical to understand that this fictional character and the real character become completely confused – there are numerous people around who believes that this person is actually fiction instead of realising that he's a fiction that *he's* created..." (PC)

Here Sinclair emphasises the permeable movement between fiction and reality in terms of his, and his characters' identity. In practice Sinclair creates a character, which may be inspired by a real life person, a historical character, or by an aspect of himself. Once created, these characters have a life of their own, and if we take this seriously, as Sinclair does, they can in turn write their own biography in the same way that we can. These imagined (alter) egos thus become as real as we perceive our own material self to be. Or at least as real as any other character in the fictive worlds of Sinclair, as he describes in relation to his own alter ego Norton:

> "'There are periods,' I said, 'when I'm not myself. I am, in fact, on a part-time basis, a fictional character in the novels of a man called Norton. I enact whatever he imposes upon me. The man has little imagination but, like a vampire, he makes steady withdrawals from my vital essence. His romances are vulgarised versions of my autobiography. I inspire, he betrays'." (Sinclair, 2001:210)

Here Sinclair again bitingly satirises his own practice for diving into his own past for characters and copy. In the spirit of multiple and (con)fusing identities, I say to Sinclair that part of me (the part that is conditioned into expecting a conventional fiction, a novel with clear characters and plotline, see Box 12.2) wants to know whether Dryfield, Silverfish or Norton are 'real' in the traditional sense; however another aspect of my 'self' (the part that knows such conventions are major simplifications of a complex, entangled world) enjoys the ambiguity and uncertainty generated by this plurality and multiplicity. He replies:

> "You can't pin these characters down [as either wholly real or wholly fictional], you just can't do it, because they become themselves and everybody is writing themselves as well as you trying to interpret them, neither of them are completely accurate. There's a hybrid form made [by them and by me as an author] and once it's published other people can read that and it adapts again and again and again. It just goes on Xeroxing itself ... and if it works you've created a mythology." (PC)

In *Landor's Tower* Sinclair has therefore created not a modern universe, but a postmodern 'multiverse'. As he writes, the 'multiverse' is the "truest model for the fix I was in… [a reality] in which alternate selves lived different lives, simultaneously" (Sinclair, 2001:105). The borderland between England and Wales functioned as the locus in which his English and Welsh selves could meet and grapple with one another.[6] A key portal into this borderland space was the crossing of the river Severn, at Aust.

The non-place of Aust: a portal into the permeable borderland

As we have seen in Chapter 5, the notion of 'non-place' alludes to homogenised locations which reduce human interactions to contractual engagements rather than meaningful relations. Allied to this however, non-places also offer a freedom in which new lives can be lived and new selves can be formed. Sinclair employs a non-place as the entry point into his multiverse, offering his characters the opportunity to re-fresh themselves before entering or exiting the borderland world. Cartographically, this non-place is located at Aust, in a motorway service station on the shores of the river Severn (see Figure 12.1). For the pod-encased character-authors of *Landor's Tower*, Aust services offers a non-place in which they could get out of their cars and encounter one another. In Sinclair's terms Aust was

---

6 It is no coincidence that the Ewyas valley "was named from the Old Welsh word 'gwyas', which meant 'place of battle'" (Sinclair, 2001:278). It was the optimal location for the encounter and conflict between the different aspects of Sinclair's identity.

the "frontier post" (2001:100), a "neutral territory between worlds" (ibid.:107). As he writes, Aust was:

> "a pleasure palace from the early days of motorway utopianism [but] had been reimagined as a zone of disappearances. I was visiting Aust at the optimum moment, when it was drifting between function and fantasy." (Sinclair, 2001:100)

> "3 a.m. in a motorway service station is where you'll find [my character-authors]. ...They all washed up here." (Sinclair, 2001:107)

At Aust the plotlines between fiction and reality did not exist, as if in a drug altered state, here:

> "material reality [is turned] into a soap opera. You become a part of whatever you're watching. You interface with EastEnders. You can actually affect the plot. ...
>
> ... 'I've got to tell you, man,' Mutton yelled at my retreating back, 'you're copping out. ...Your prose is getting really slack, all those one-word sentences, the reliance on a narrow band of imprecise adjectives. Neglect the armature of grammar and the world loses definition'.[7]
>
> ... I needed a walk to clear my head after my encounter with the Ketamine Kreeps. ...Now, as always, there was the pom-pom-pom of traffic speeding out of Wales." (Sinclair, 2001:109)

Box 12.2 Puppet masters and uncontrolled narratives: Iain Sinclair and Barbara Vine

As we have seen, Iain Sinclair's psychogeographical style subverts the chronological and linear narrative structure associated with 'modern' novels. His characters are amalgams of the real and imagined, and his stories do not conform to a traditional triptych of 'beginning',

---

7 Here again a version of Sinclair is ridiculing his own psychogeographic prose style. His neglect for grammar and penchant for one word sentences is illustrated in the following excerpt from 'London Orbital': "Autumn. No hat. A war footing: mufti-awkward. Argie bashing, ranting. Cromwell-fierce, hormonally stoked, she wields her small scythe, dismissing the unseen enemy, stalkers in the bushes, eco-bandits, twitchers, pennypinchers, lily-livered Liberal fifth-columnists, bedwetters, nay-sayers" (2002:3-4).

'middle' and 'end'. In *Landor's Tower,* Sinclair takes one traditionalist novelist to task over the linearity and predictability of her novels' form. I ask Sinclair about the discussion of Barbara Vine in his plot. I say that the tone of his remarks suggests that it is almost as if you – or Norton (or both?) – were quite jealous of the ability of modern novelists to simplify their fiction, limiting as they do so the complexities associated with more postmodern writing like his own.

> "There was the kind of humour of that, where I was literally reading this book and seeing how extraordinarily different the set of procedures was to what I was doing. ...This is what we are told to do if you were supposed to be a professional writer but it's controlled and it's a world where the writer as puppeteer is too visible. Even though supposedly there's no Barbara Vine [as she is a pseudonym of Ruth Rendell] the presence of the person doing it is even stronger ironically than one of these novels that the writer is in it and not in it and gone, and that's hard to get hold of. It's as if you sign up to go on a journey, you know exactly what the journey is and if she subverted it and did something terrible then everybody would scream.... This works well on certain terms but why for me it was finally not satisfying because it's such a simple equation in the end that has to be resolved; whereas the world I want to present is this democratic thing that doesn't have to be resolved and you don't quite know what you are signing up for in some of these other forms, sometimes they work out well and sometimes they're disastrous..." (PC)

As he writes in *Landor's Tower*:

> "My pleasure in the deftness of her performance was double-edged: I relished the parallel world to which I had been granted access, but I was unnerved by the way she did it. This was how fiction intended for an audience should be assembled; the tiny detail in Chapter 3 that will be resolved in Chapter 21; past events remembered in present time, teases, hooks, cross-cuts far slicker than the standard soap opera. And, always, beneath the nicely managed surface, cruelty and pain. The impossibility of human relationships. Karma. Ancient crimes, insignificant acts that blight future ease. A man deciding, out of character, to walk down an icy path. A woman who doesn't pick up a ringing phone. Nothing is here that doesn't belong here. ...The reader is nurtured, humoured, without the condescension of the grande dames of the Golden Age of mystery fiction." (Sinclair, 2001:30)

Here Sinclair implies that modern mystery fiction is too clean, easy, and predictable for the reader; the reader expects to be pacified, spot the odd red herring or two, and simply sit back and enjoy the read/ride. From Sinclair's perspective, the reader is "drawn into [the author's] world, that world possesses you: the sound of the words, your breath intoning sentences, succumbing to flow, going under" (2001:31). For Sinclair, the modern style results in the reader becoming a spectator on a journey of the novel, rather than an active participant in it; it results in the death of the reader, rather than the death of the author (see Chapter 1). This is writing and reading as a one way conversation, a monologue on behalf of the author as puppet master.

In the more post-modern style that Sinclair adopts for *Landor's Tower*, the reader's world is different. Here the characters have their own agency and lives outside the text, whilst the reader can, and must, interpret and reinterpret the journey in order to gain understanding. As Sinclair states, the reader must "interface with [this reality]. [And as a consequence, they] can actually affect the plot..." (2001:109). In Sinclair's plotline, the writer is not dead, but has little control over proceedings. Echoing themes in Chapter 7, in these plotlines even the words of the author cannot be relied upon as intentional and faithful presentations of a singular reality; as he writes:

> "This is not what I felt about Wales, not what I set out to say.
>
> You can make as many charts as you like, plot graphs with different-coloured inks, predict movements, the arguments of ungrateful characters. You can spend years ploughing through biographies, reminiscences by tourists in search of the pastoral; libraries of geology, church histories, mythology. You can visit every site a dozen times, live on the road. It makes no difference. The first sentence on the page and the game's up, the story goes its own way." (Sinclair, 2001:307)

In this genre, the author is not a puppeteer, but simply another actor in the cast, with sometimes less control than any other. As Sinclair (through the guise of Norton) admits: "One thing I had learnt, the last person you should ask for a solution is the author" (Sinclair, 2001:252). Due to this lack of control, in Sinclair's novel the plotline becomes an "endlessly revolving carousel" (PC) between (mis)understandings, approximations and (mis)representations, between all the key agents in the plot: the author, reader, author-characters and place. As Sinclair mockingly comments, in comparison to a conventionally circumscribed narrative: "The Norton novel meant nothing; a folly of the worst kind, forged narratives, bent history. A disservice to all concerned" (Sinclair, 2001:307).

Conclusion as Carousel

In the postmodern mash-up that is *Landor's Tower*, Iain Sinclair has produced a novel that blurs the boundaries between England and Wales, fiction and reality, past and present, and perhaps even sanity and madness. In its simplest form, *Landor's Tower* is a homecoming narrative, exploring the nature of exile from one's hiraeth and the reminiscences and changes one encounters when meeting again the people and place of one's youth. At his most complex, Sinclair has created his own multiverse, splicing his identity with other real and imagined characters in order to produce not only a gonzo-inspired road movie narrative, but also "a new mythology" for the area and "a cosmology" for life (PC).

In *Landor's Tower* Sinclair is thus not only exploring his own homecoming and the way our identities are increasingly fragmented by different places and times, but also the nature of reality. In terms of our identity, Sinclair argues that people have multiple selves which do not form an integrated, consistent whole; rather they form a contradictory, paradoxical and juxtaposed self. He suggests that identity is a battle ground, a contact zone, and in the borderland between England and South Wales, Sinclair pits various aspects of his identity against one another in order to 'work through a whole'. This work is the 'complication', in the words of Kondo, of "decentred, multiple selves, whose lives are shot through with contradictions and creative tensions" (1990:220).

At first glance, this conceit appears to be a simulated version of reality, a literary device used in order to forge a complex and often disorientating narrative. However, for Sinclair, this 'reality' is not the simulation. Rather this 'multiverse' is another way in which reality can be imagined and understood: in the mobile, fractured and globalised society in which we live, it is difficult to be integrated and consistent in terms of our identity because of the disparate forces pulling relations in different directions. In his view, we are living in a world where boundary crossing of all kinds is increasingly prevalent (see Reed-Danahay, 1997:3), and thus identity (be it of self or place) is becoming an arena in which "disparate [entities/processes] meet, clash, and grapple with each other" (after Pratt, in Warren, 1997:4). From this perspective, to be inconsistent and inauthentic is the nature of things, but to have the appearance of consistency and authenticity remains the aim. In many senses, therefore, the more the world and its people seem specific and essential, the more likely they are to be fake. In this view, the modern and simplified (be it in terms of mystery novels or singular places) is the simulacra, only the plural multiverse is 'real'.

This post-modern way of conceiving reality therefore suggests that there is no inherent order in the world. As Whatmore writes (1999:31-2), order is "always in the making, not... a priori". Thus Sinclair's 'endlessly revolving carousel' of a novel, without a beginning and end or three part structure, is in fact more closely reflective of a postmodern reality than one may initially think. Ironically perhaps, the traditionally structured fiction is the simulacra, the attempt to pass off reality with a fake version that attempts to fool or passify us (see Baudrillard, 1994). Sinclair's bewilderingly puzzling narrative with no (re)solution that itself is an entanglement of real and imagined and page and place, is a truer approximation of the world than we first envisaged.

Sinclair therefore tells us that we should not aim for conclusion or closure, but for a "permanently transitional status" (2001:244). Just as there is no singular plotline, or one means to understand a narrative, there is rather only the view from where you are, and that place itself changes with time and trajectory. In line with the argument of *Page and Place* there is always a plurality of realities as we live in a world of infinite places (after Solnit, 2010). In one sense, the entanglement of page and place is a puzzle with no solution, but nevertheless through the process of exploring plotlines we can gain a fresh understanding about the human condition and our relations to the places that define us.

Sinclair reminds us that despite this world of constant change our relations to geography remain vital for our sense of identity. As he states: "every plot has to have a believable landscape, a map you can trust" (2001:99). This map could be one which clearly distinguishes rivers from roads and provides 'near factual statistical documentation', or it could be a map which entangles the real with the imagined to give a more oblique and poetic understanding of the times and places in which we build our lives. In opting for an entangled map of page and place, Sinclair (and this book) has explored how the lines between fiction and geography are not a clear, unimpeachable border, but rather "a dissolving line between sanity and madness; genius and gibberish; legend and reality" (Adams, 2001:no page). Although often bewildering and disorientating, this map enables us to gain important insights in the changing nature of the human (geographical) condition. As Sinclair states:

> "the big point is I think is how we become defined as human entities by place and memory. We bounce off known objects and aspects of landscape as we move through it so we know ourselves by that process. If you go out into that and its not there then parts of your own memory and your essential being, how you define yourself, are not there either, so it's very challenging and people become completely disturbed and sometimes they don't know why they are but that is the central reason for it. [Without connection to the land and our memories a person] ceases to be himself [sic], he doesn't have the markers, he doesn't know who he is so he becomes a ghost of his own history..." (PC)

Here Sinclair may be referring directly to his Hackney neighbours who have become displaced by transnational-led redevelopment, or the process of his return to Wales, but his point remains a vital touchstone in our own constitutive coingredience (Casey, 2001) with the places in which we have lived and continue to live.

Thus as we have seen in this chapter, and more extensively in this book, literature plays a key role in helping us to understand and belong in our geographies. Literature contributes to a "layering of place [that] is so amazing and exciting [that our] story accumulates..." (Cooke, 2009), involving a juxtaposition of different character-authors and fictions that produce a "mangled histor[y]" and entangled geography which is a "labyrinthine fiction..." (Norton, in Sinclair, 2001: 174) but is nevertheless true. Using literature can therefore help us to democratise knowledge away from dominant voices and broaden cartography away from simple map fetishism. Fiction, in short, can help us tell a new truth about the relations between people and place. It can move us towards an understanding of the existence of plural plotlines and the ways in which these are formed by entanglements of page and place. From these ongoing compositions we can locate our own lives, finding new meanings which resonate with our own experiences and broaden our horizons, accepting the inevitable tensions between being rooted and mobile, insular and outward facing. In other words, it can remind us to celebrate the entanglements which exist between page and place.

Bibliography

Abbey, Edward (1975): *Desert Solitaire A Season in the Wilderness*. New York: Ballantine Books.

——— (1991): *The Monkey Wrench Gang*. London: Robin Clark.

Abley, Mark (2003): *Spoken Here. Travels Among Threatened Languages*. London: Arrow.

Ackroyd, Peter (2000): *London: The Biography*. London: Chatto and Windus.

——— (2007): *Thames: Sacred River*. London: Chatto and Windus.

Adams, Tim (2001): "Singing His Prose." *Observer. The Guardian.com*. 8 April 2001. <http://www.guardian.co.uk/books/(2001)/apr/08/fiction.features>.

Agnew, John and James Duncan, eds. (1989): *The Power of Place: Bringing Together Geographical and Sociological Imaginations*. London: Unwin Hyman.

Aitken, Stuart C. and Deborah P. Dixon (2006): "Imagining Geographies of Film." *Erdkunde* 60: 326-36.

Alexander, Neal (2010): *Ciaran Carson. Space, Place, Writing*. Liverpool: Liverpool University Press.

Allen, Michael (2006): "Malcolm Pryce: The Unbearable Lightness of Being in Aberystwyth." *Grumpy Old Bookman.Blogspot.co.uk*. 7 Jan. 2006 <http://grumpyoldbookman.blogspot.co.uk/(2006)/01/malcolm-pryce-unbearable-lightness-of.html>.

Álvarez, David (2009): "'Dis Poem is Vex About Apartheid': Representations of South Africa in Three West Indian Poems." *Journal of Commonwealth Literature* 44.2: 107-22.

Anderson, Benedict (2006): *Imagined Communities: Reflections on the Origin and Spread of Nationalism*. London: Verso.

Anderson, Jon (2002): "Researching Environmental Resistance: Working Through Secondspace and Thirdspace Approaches." *Qualitative Research* 2.3: 301-21.

——— (2004a): "Talking Whilst Walking: A Geographical Archaeology of Knowledge." *Area 36*.3: 254-61.

——— (2004b): “Spatial Politics in Practice, The Style and Substance of Environmental Direct Action.” *Qualitative Research* 2.3: 301-21.

——— (2004c): “The Ties that Bind? Self- and Place-Identity in Environmental Direct Action.” *Ethics, Place and Environment* 45-58.

——— and Kate Moles (2008): “Walking into Coincident Places.” *Qualitative Researcher* 9: 5-7.

——— (2009): “Transient Convergence and Relational Sensibility: Beyond the Modern Constitution of Nature.” *Emotion, Space & Society* 2: 120-27.

——— (2010): *Understanding Cultural Geography: Places and Traces.* London and New York: Routledge.

——— (2014): “What I Talk About When I Talk About Kayaking.” *Water Worlds. Human Geographies of the Ocean.* Eds. Jon Anderson and Kimberley Peters. Farnham: Ashgate. 103-18.

Anzaldua, Gloria (1987): *Borderlands: La Frontera.* San Francisco: Aunt Lute Books.

Arefi, Mahyar (1999): “Non-Place and Placelessness as Narratives of Loss: Rethinking the Notion of Place.” *Journal of Urban Design* 4.2: 179-93.

Atkinson, Kate (1995): *Behind the Scenes at the Museum.* London: Black Swan.

Augé, Marc (1995): *Non Places: Introduction to an Anthropology of Supermodernity.* Trans. John Howe. London: Verso.

Bairner, Alan (2011): “Urban Walking and the Pedagogies of the Street.” *Sport, Education and Society* 16.3: 371-84.

Barthes, Roland (1973) *Mythologies.* London: Paladin.

Baudrillard, Jean (1981): *For a Critique of the Political Economy of the Sign.* St. Louis: Telos.

——— (1994): *Simulacra and Simulation.* Michigan: University of Michigan Press.

Bauman, Zigmund (2000): *Liquid Modernity*, Cambridge: Polity Press.

——— (2002): “Foreword. Individually, Together.” Beck and Beck-Gernsheim 2002: XIII-XX.

——— (2003): *Liquid Love: On the Frailty of Human Bonds,* Cambridge: Polity Press.

——— (2005): *Liquid Life*. Cambridge: Polity Press.

BBC (2012): "Vulcan Pub in Cardiff Dismantled for St Fagans Museum Rebuild in 'Several Years." 4 May 2012. <http:// www.bbc.co.uk/news/uk-wales-south-east-wales179552 68>.

——— (2012): "Aber Mon Amour." 23 Apr. 2012. <http://downloads. bbc.co.uk/wales/archive/bbc-mid-wales-boo ks-malcolm-pryce-interview.pdf>.

Beck, Ulrich (1992): *Risk Society: Towards a New Modernity*. London: Sage.

——— (1994): "The Reinvention of Politics: Towards a Theory of Reflexive Modernisation." *Reflexive Modernization: Politics, Tradition and Aesthetics in the Modern Social Order*. Eds. Ulrich Beck, Anthony Giddens and Scott Lash. Cambridge: Polity Press. 3-23.

——— (1995): *Ecological Politics in an Age of Risk*. Cambridge: Polity Press.

Belk, Russell W. (1990): "The Role of Possessions in Constructing and Maintaining a Sense of the Past." *Advances in Consumer Research* 17: 669-76.

Benjamin, Walter (2002): *The Arcades Project*. Harvard: Harvard University Press.

Bey, Hakim (1991): *T.A.Z. The Temporary Autonomous Zone: Ontological Anarchy and Poetic Terrorism.* New York: Autonomedia.

Bhabha, Homi (1994): *The Location of Culture*. London: Routledge.

Blair, Caroline, Greg Dickinson and Brian L. Ott. (2010): "Introduction: Rhetoric/Memory/Place." *Places of Public Memory: The Rhetoric of Museums and Memorials*. Eds. Caroline Blair, Greg Dickinson and Brian L. Ott. Tuscaloosa: University of Alabama Press. 1-54.

Blake, William (1863): *Auguries of Innocence.* London: Ruthven Todd.

Bonnett, Alastair (2009): "The Dilemmas of Radical Nostalgia in British Psychogeography." *Theory, Culture, Society* 25.1: 45-70.

Botton, Alain de (2002): *The Art of Travel.* London: Hamish Hamilton.

Brenner, Joseph, Robert Coles and Dermot Meagher (1970): *Drugs and Youth.* New York: W.W. Norton.

Brockway, Antony (2004): "An Interview with Niall Griffiths." <http://homepage.ntlworld.com/elizabeth.ercocklly/niall.htm>.

Brooks, Paul (1980): *Speaking for Nature: How Literary Naturalists from Henry Thoreau to Rachel Carson have Shaped America.* Boston: Houghton Mifflen.

Bryden, Inga (2012): "There are Different Ways of Making the Streets 'Tell'. Narrative, Urban Space and Orientation." *Writing the Modern City. Literature, architecture, modernity.* Eds. Sarah Edwards and Jonathan Charley. London: Routledge. 213-26.

Buttimer, Anne (1976): "Grasping the Dynamism of Lifeworld." *Annals of the Association of American Geographers* 66.2: 277-92.

Burgess, Jackie and John Gold (1985): *Geography, the Media and Popular Culture*. London: Crook Helm.

Burke, Edmund. (1761): *A Philosophical Enquiry into the Origin of Our Ideas of the Sublime and Beautiful*. London: RJ Dodsley.

Calvino, Italo (1986): *Invisible Cities*. London: Martin Secker and Warburg.

Casey, Edward (2000): *Remembering. A Phenomenological Study.* Bloomington: Indiana University Press.

——— (2001): "Between Geography and Philosophy: What Does It Mean to Be in the Place-World?" *Annals of the Association of American Geographers* 91.4: 683-93.

——— (2005): "Smooth Spaces and Rough-Edged Places: The Hidden History of Place." *Review of Metaphysics* 51.2: 267-96.

Chatwin, Bruce (1987): *Songlines*. London: Vintage.

Clare, John (1908): "Poems by John Clare." Ed. Arthur Symons. *Archive.org.* <http://www.archive.org/stream/poemsbyjohnclare00clariala/poemsbyjohnclare00clariala_djvu.txt>.

Clarke, Gillian (1971): *Snow on the Mountain*. Aberystwyth: Christopher Davies.

——— (1978): *The Sundial*. Llandysul: Gomer.

——— (1982): *Letter from a Far Country*. Manchester: Carcanet Press.

——— (1997): *Collected Poems*. Manchester: Carcanet Press.

——— (1998): *Five Fields*. Manchester: Carcanet Press.

——— (2003): "City of Tall Stories." *Cardiff Central. Ten Writers Return to the Welsh Capital*. Ed. Dannie Abse. Llandysul: Gomer. 51-64.

——— (2008): *At the Source*. Manchester: Carcanet Press.

Clifford, Nick and Gill Valentine, Eds. (2003): *Key Methods in Geography*. London: Sage.

Coleman, Simon and Mike Crang (2002): "Grounded Tourists, Travelling Theory." *Tourism: Between Place and Performance*. Eds. Simon Coleman and Mike Crang. New York: Berghahn Books. 1-17.

Collins, Richard (2004): *The Land as Viewed from the Sea*. Bridgend: Seren.

——— (2006): *Overland*. Bridgend: Seren.

——— (2011): *The Quality of Light*. Bridgend: Seren.

Community Cohesion Panel, (2004): *The End of Parallel Lives? The Report of the Community Cohesion Panel*. London: Home Office.

Community Cohesion Review Team (2001): *Community Cohesion: A Report of the Independent Review Team*. London: Home Office.

Convery, Ian et al. (2005): "Death in the Wrong Place? Emotional Geographies of the UK 2001 Foot and Mouth Disease Epidemic. *Journal of Rural Studies*. 21.1: 99-109.

Cook, Ian and Mike Crang (1995): *Concepts & Techniques in Modern Geography: Doing Ethnographies*. Durham: CATMOGS.

Cook, Ian (1997): "Participant Observation." Flowerdew & Martin 1997: 127-50.

Cooke, Rachel (2009): "The Interview: Iain Sinclair." *Observer.The Guardian.com*. 8 Feb. 2009 <http://www.theguardian.com/ books /2009/feb/08/iain-sinclair-interview>.

Cooper David and Ian N. Gregory (2011): "Mapping the English Lake District: A Literary GIS." *Transactions of the Institute of British Geographers* 36.1: 89-108.

Cosgrove, Daniel (1989): "Geography is Everywhere: Culture and Symbolism in Human Landscapes." *Horizons in Human Geography*. Eds. Derek Gregory and Rex Walford. Basingstoke: Macmillan. 118-35.

Coverley, Merlin (2006): *Psychogeography*. Harpenden: Pocket Essentials.

Crang, Mike (1998): *Cultural geography*. London: Routledge.

——— (2003): "Qualitative Methods: Touchy, Feely Look-See?" *Progress in Human Geography* 27.4: 494-504.

——— and Penny Travlou (2001): "The City and Topologies of Memory." *Environment & Planning D: Society & Space* 19.2: 161-77.

Crawshaw, Robert and Corinne Fowler (2008): "Articulation, Imagined Space and Virtual Mobility in Literary Narratives of Migration." *Mobilities* 3.3: 455-69.

Cresswell, Tim (1996): *In Place/Out of Place*. Minneapolis: University of Minnesota Press.

——— (2004): *Place: A Short Introduction*. Oxford: Blackwell.

Darwin, Charles (1996): *Natural Selection*. London: Phoenix.

Davidson, Joyce (2003): *Phobic Geographies. The Phenomenology of Spatiality and Identity*. Aldershot: Ashgate.

———, Liz Bondi and Mick Smith (2005): *Emotional Geographies*. Aldershot: Ashgate.

Davies, Grahame (2004): *Rhaid i Bopeth Newid*. Llandysul: Gomer.

——— (2005): "Grey." Finch & Davies 2005: 46.

——— (2005): "Number 62 Bus." Finch & Davies 2005: 47.

——— (2007): *Everything Must Change*. Bridgend: Seren.

Davies, John (1994): *The History of Wales*. London: Penguin.

de Certeau, Michel (1984): *The Practice of Everyday Life*. Berkley: University of California Press.

Delanda, Manuel (2006): *A New Philosophy of Society. Assemblage Theory and Social Complexity.* London: Continuum.

Deleuze, Gilles (1985): "Nomad Thought." *The New Nietzsche*. Ed. David Allison. Cambridge: MIT Press. 142-149.

——— (1993): *The Fold.* Minneapolis: University of Minnesota Press.

———and Guattari Felix (1981): *A Thousand Plateaux: Capitalism & Schizophrenia.* Minneapolis: University of Minnesota Press.

Devall, Bill and George Sessions (1985): *Deep Ecology: Living as Nature Mattered.* Layton: Gibbs M Smith.

Dicken, Peter (2003): "*Global shift*: *Reshaping the Global Economic Map in the 21st Century.*" London: Sage.

Dillard, Annie (1975): *Pilgrim at Tinker Creek*. London: Picador.

——— (1982): *Teaching a Stone to Talk*. London: Picador.

——— (1999): *For the Time Being*. London: Picador.

Doel, Marcus (1999): *Poststructuralist Geographies: The Diabolical Art of Spatial Science*. Lanham: Rowman & Littlefield.

Dorst, John (2013): "Skin Remembers: Animal Trophies as Material Memory Cultural Studies," *Critical Methodologies* 13.1: 35-46.

Dovey, Kim (2010): *Becoming Places. Urbanism/Architecture/ Identity/Power.* Abingdon: Routledge.

Ehrlich, Gretel (1985): *The Solace of Open Spaces*. New York: Penguin.

Elfyn, Menna, Ted Hughes and Seamus Heaney (1997): *Trying the Line. A Volume of Tribute to Gillian Clarke.* Llandysul: Gomer.

Ellis, Caroline (2004): *The Ethnographic I: A Methodological Novel about Ethnography*. Alta Mira: Walnut Creek.

Elmes, Simon (2005): *Talking for Britain. A Journey Through the Nation's Dialects*. London: Penguin.

Entrikin, Nicholas J. (2001): "Hiding Places." *Annals of the Association of American Geographers* 91.4: 694-7.

Evans, Rhrodi (2013): "A Welsh Landscape." *TheCuriousAstronomer.Wordpress.com.* 21 Feb. 2013 <http://thecuriousastro

nomer.wordpress.com/(2013)/02/21/a-welsh-landscape-r-s-thomas-poem/>.

Faulkner, William (1996): *As I Lay Dying*. London: Vintage.

Featherstone, Mike (1995): *Undoing Culture*. London: Sage.

Finch, Peter (1987): *Selected Poems*. Bridgend: Poetry Wales Press.

——— (2002): *Real Cardiff.* Bridgend: Seren.

——— (2003): "Cardiff – The City that Floats." *Cardiff Central. Ten Writers Return to the Welsh Capital*. Ed. Dannie Abse. Llandysul: Gomer. 13-26

——— (2004): *Real Cardiff 2*. Bridgend: Seren.

——— (2005): "St Mary Street." *Finch & Davies 2005*. 79.

——— (2007): *Selected Later Poems*. Bridgend: Seren.

——— (2008): *Real Wales*. Bridgend: Seren.

——— (2009): *Real Cardiff 3*. Bridgend: Seren.

———and Grahame Davies Eds. (2005): *The Big Book of Cardiff.* Bridgend: Seren.

——— Ian Davidson and Zoe Skoulding (2006): "Peter Finch Interview." *ArgotistOnline.co.uk.* <http://www.argotistonline.co.uk/Finch%20interview.htm>.

Fisher, Allen (2005): *PLACE*. Hastings: Reality Street Editions.

Flowerdew, Robin and David Martin (1997): *Methods in human geography.* n.p. Longman.

Forshaw, Barry (2001): Good Reads. <http://www.goodreads.com/book/show/398081.Aberystwyth_Mon_Amour>.

Foucault, Michel (1979): *Discipline and Punish: The Birth of a Prison*. Harmondsworth, Middx: Penguin Books.

——— (1980): *Power/Knowledge*. New York: Pantheon.

Fox, Warwich (1990): *Towards a Transpersonal Ecology. Developing New Foundations for Environmentalism*. Boston: Shambala.

Garton Ash, Tim (2005): *Free World*. London: Penguin.

Gleick, James (1990): *Nature's Chaos*. New York: Viking.

Gottlieb, Robin (1991): *This Sacred Earth. Religion, Nature, Environment*. New York: Routledge.

Griffiths, Lloyd (2011): "Owen Sheers. Resistance Q&A." *Buzz Magazine.co.uk*. 10 Nov. 2011 <http://www.buzzmag.co.uk/uncategorized/owen-sheers-resistance-qa/>.

Griffiths, Niall (2000): *Grits*. London: Vintage.

——— (2001): *Sheepshagger*. London: Vintage.

——— (2003): *Stump*. London: Vintage.

——— (2007): *Runt*. London: Vintage.

——— (2010): *The Dreams of Max and Ronnie*. Bridgend: Seren.

——— (2008): *Real Aberystwyth*. Bridgend: Seren.

Gupta, Anil (2011): Resistance. Big Rich Films.

Gustafson, Per (2001): "Roots and Routes – Exploring the Relationship Between Place Attachment and Mobility." *Environment & Behavior*. 33.5: 667-86.

Hadley, Tessa (2005): "The Trouble with Summer." Finch & Davies 2005. 91-96.

Hadley, Tessa (2012): *The London Train*. London: Vintage.

Halbwachs, Maurice (1992): *On Collective Memory*. Chicago: University of Chicago Press.

Hall, Stuart, Ed. (1997): *Representation: Cultural Representations and Signifying Practices*. London: Sage.

Hammersley, Mikey and Paul Atkinson (1995): *Ethnographv: Principles into Practice*. London: Routledge.

Harrison, Fraser (1986): *The Living Landscape*. London: Mandarin.

Harvey, David (1996): *Science, Nature and the Geography of Difference*. Oxford: Blackwell.

Heaney, Seamus (1966): *Death of a Naturalist*. London: Faber & Faber.

——— (1980): *Preoccupations. Selected Prose 1968-1978*. London: Faber & Faber.

Heartfield, James (2004): "Londonnostalgia." *Heartfield.org*. Blueprint, Sept. 2004. <http://www.heartfield.org/londonostalgia.htm>.

Heidegger, Martin (1956): *Being and Time*. Oxford: Blackwell.

Henry, Graham (2013): "Welsh Founder of Hughesovka Honoured by Ukraine With Museum and Commemorative Stamp. *Wales Online.co.uk.* 11 Nov. 2013. <http://www.walesonline.co.uk/news/wales-news/welsh-founder-hughesovka-honoured-ukraine-6289244>.

Herbert, Steve (2000): "For Ethnography." *Progress in Human Geography* 24.4: 550-68.

Hillier, Jean (2001): "Imagined Value: The Poetics and Politics of Place." *People and Place: The Extraordinary Geographies of Everyday Life*. Eds. Ali Madanipour, Lewis Holloway and Phil Hubbard. New York: Prentice Hall.

Ho, E. (2006): "Postimperial Landscapes: "Psychogeography" and Englishness in Alan Moore's Graphic Novel From Hell: A Melodrama in Sixteen Parts" *Cultural Critique* 63: 99-121.

Holm, Douglas K. (2001): "Landor's Tower." *Powells.com.* 29 Oct. 2001 <http://www.powells.com/review/(2001)_10_29.html>.

Home, Stewart (1997): "Introduction – Mondo Mythopoesis" *Mind Invaders: A Reader in Psychic Warfare, Cultural Sabotage and Semiotic Terrorism*. Ed. Stewart Home. London: Serpent's Tail.

Hones, Sheila (2008): "Text as It Happens: Literary Geography." *Geography Compass* 2.5: 1301-17.

hooks, bell (1991): *Yearning: Race, Gender and Cultural Politics.* Boston: South End Press.

——— (2008): *Belonging: A Culture of Place*. New York: Routledge.

Hornsey, Richard (2012): "Listening to the Tube Map: Rhythm and the Historiography of Urban Map Use" *Environment and Planning D: Society and Space* 30.4: 675-93.

Ianto's Shrine. (2013): "Ianto's Shrine, Mermaid Quay, Cardiff Bay." *Foursquare.com* <https://foursquare.com/v/iantos-shrine/4cbb007b4495721eff2f537a>.

Illych, Ivan (1973): *Tools for Conviviality*. London: Fontana.

Ingold, Tim (1993): "The Temporality of the Landscape." *World Archaeology*. 25.2: 152-74.

——— (2005): "Up, Across and Along." *SpaceSyntax.TUDelft.nl* 13-17 June 2005. <http://www.spacesyntax.tudelft.nl/media/Long%20papers%20I/tim%20ingold.pdf>.

Jameson, Fredric (1991): *Postmodernism or the Cultural Logic of Late Capitalism*. London: Verso.

Johnson, Nuala (2004): "Fictional Journeys: Paper Landscapes, Tourist Trails and Dublin's Literary Texts." *Social & Cultural Geography* 5.1: 91-107.

Jones, Alice (2011): "A Writer Who's Hard to Resist." *TheIndependent.co.uk* 24 Nov. 2011 <http://www.independent.co.uk/news/people/profiles/a-writer-whos-hard-to-resist-6267092.html>.

Jones, Katie (2007): *Narratives of the In-Between: Teenagers' Identities and Spatialities in a North Wales Town*. Diss. Cardiff University.

Jones, Martin (2009): "Phase Space: Geography, Relational Thinking, and Beyond. *Progress in Human Geography* 33.4: 1-20.

Jordan, Justine (2001): "Last Exit to Aberystwyth." *TheGuardian.com* 3 Feb. 2001 <http://www.guardian.co.uk/books/(2001)/feb/03/fiction.reviews2>.

Keiller, Patrick (1994): "London" Koninck/Channel 4.

——— (1997): "Robinson in Space." BBC/Koninck.

Keith, Michael and Steve Pile, Eds. (1993): *Place and the Politics of Identity*. London: Routledge.

Kerouac, Jack (1999): *On the Road*. Harmondsworth: Penguin.

Kitchen, Rob and Nicholas Tate (2000): *Conducting Research in Human Geography: Theory, Methodology and Practice*. Harlow: Prentice Hall.

——— and Martin Dodge (2007: 341): "Rethinking maps" cited in Hornsey 2012.

Knights, Ben and Chris Thurgar-Dawson (2011): *Active Reading: Transformative Writing in Literary Studies*. London: Continuum.

Kondo, D. (1990): *Crafting Selves: Power, Gender and Discourses of Identity in a Japanese Workplace*, Chicago: University of Chicago Press.

Koppel, Gideon (2008): "Sleep Furiously." New Wave Films.

Kropotkin, Peter (1915): *Mutual Aid: A Factor of Evolution*. London: Heinemann.

Kusenbach, Margarethe (2003): “Street Phenomenology: The Go-Along as Ethnographic Research Tool.” *Ethnography* 4.3: 455-85.

Kymlicka, Will (2003): “Multicultural States and Intercultural Citizens.” *Theory and Research in Education* 1.2: 147-69.

Lahaie, Christiane (2008): “Between Geography and Literature - The Interplay of Place and Mimesis.” *Cahiers de Geographie du Quebec* 52.147: 439-51.

Latour, Bruno (1993): *We Have Never Been Modern*. Cambridge: Harvard University Press.

Lee, T. (2010): Streets of Cardiff. <http://www.bbc.co.uk/wales/southeast/sites/streetsofcardiff/pages/redhouse_terrylee.shtml>

Leopold, Aldo (1949): *A Sand County Almanac and Sketches From Here and There.* London: Oxford University Press.

Leopold, Luna, ed. (1953): *From the Journals of Aldo Leopold: Round River*. New York: Oxford University Press.

Ljungberg (2003: 174): “Constructing New ‘‘Realities’’: The Performative Function of Maps in Contemporary Fiction”, also cited in Piatti & Hurni, 2009: 340.

Llewellyn, Richard (2001): [1939] *How Green was my Valley*. London: Penguin.

Lopez, Barry (1992): *The Rediscovery of North America*. London and New York: Vintage.

Lopez, Barry (1998): About this Life. London: Harvill Press.

——— (2001): *Arctic Dreams*. London and New York: Vintage.

——— (2004): *Field Notes*. London and New York: Vintage.

——— and Debra Gwartney, Eds. (2006): *Home Ground. Language for an American Landscape*. San Antonio: Trinity.

Lossau, Julia and Roland Lippuner (2004): “Geography and the Spatial Turn.” *Erdkunde* 58.3: 201-11.

Low, Setha M. (1992): “Symbolic Ties that Bind. Place Attachment in the Plaza.” *Place Attachment. Human Behavior and Environment. Advances in Theory & Research. Volume 2.* Eds. Irwin Altman and Setha M. Low. New York & London: Plenum Press. 165-185.

Macauley, David (2000): "Walking the City: An Essay on Peripatetic Practices and Politics." *Capitalism, Nature, Socialism* 11.4: 3-43.

Mack, John (2011): *The Sea. A Cultural History.* Chicago: University of Chicago Press.

Maffesoli, Michel (1996): *The time of the tribes*. London: Sage.

Manes, Chris (1990): *Green Rage: Radical Environmentalism in the Unmaking of Civilisation*. Boston: Little, Brown & Co.

Manzo, Lynne C. (2003): "Beyond House and Haven: Toward a Revisioning of Emotional Relationships with Places." *Journal of Environmental Psychology* 23: 47–61.

Massey, Doreen (1993): *Space, Place and Gender*. Cambridge: Polity Press.

——— (1995): "Thinking Radical Democracy Spatially." *Environment and Planning D, Society and Space* 13.3: 283-8.

——— (2005). *For Space*. London: Sage.

——— (2006a): "Space, Time and Political Responsibility in the Midst of Global Inequality." *Erkunde* 60.2: 89-95.

——— (2006b): "Landscape as a Provocation. Reflections on Moving Mountains." *Journal of Material Culture* 11.1-2: 33-48.

Matthews, Hugh et al. (2000): "The Unacceptable Flâneur: The Shopping Mall as a Teenage Hangout." *Childhood* 7.3: 279-94.

McKay, George (1998): *DiY Culture. Party and Protest in Nineties Britain*. London: Verso.

Merleau-Ponty, Maurice (1962): *Phenomenology of Perception*. London: Routledge & Kegan Paul.

Miller, Daniel (1998): *A Theory of Shopping*. Oxford: Polity Press.

Mitchell, David (2001): *Number9Dream.* London: Hodder & Stoughton.

Moorcock, Michael (1988): *Mother London*. London: Secker and Warburg.

——— (2001): "Landor's Tower; Or, The Imaginary Conversations." *RevolutionSF.com* 9 Sept. 2001. <http://www.revolutionsf.com/article.php?id=431>.

Morris, Brian (2004): "What We Talk About When We Talk About 'Walking in the City'." *Cultural Studies* 18.5: 675-97.

Muir, John (2004): *My First Summer in the Sierra.* New York: Dover.

Murdoch, Jonathan (2005): *Post-structural Geography: A Guide to Relational Space*. Thousand Oaks: Sage.

Naess, Arne (1989): *Ecology, Community & Lifestyle. Outline of an Ecosophy*. Cambridge: Cambridge University Press.

O'Loughlin, John and Luc Anselin. (1992). "Geography of International Conflict and Cooperation: Theory and Methods" *The New Geopolitics* Ed. Michael D. Ward. Philadelphia: Gordon and Breach.

Oakes, Timothy (1997): "Place and the Paradox of Modernity." *Annals of the Association of American Geographers* 87.3: 509-31.

——— and Patricia Price, Eds. (2008): *The Cultural Geography Reader*. London and New York: Routledge.

Okely, Judith and Hellen Callaway Eds. (1992): *Anthropology and Autobiography*. (ASA Monographs 29). London and New York: Routledge.

Owen, J. (2011). The Day Aberystwyth Stood Still, by Malcolm Pryce. July, 20. <http://www.bookgeeks.co.uk/(2011)/07/20/the-day-aberystwyth-stood-still-by-malcolm-pryce/>.

Peach, Linden (1993): *Ancestral Lines.* Bridgend: Seren.

Pearson, Mike and Michael Shanks (2001): *Theatre / Archaeology.* London and New York: Routledge.

Penman, Ian (2001): "A Magus Marooned." *Observer.TheGuardian .com.* 5 May 2001. <http://www.guardian.co.uk/books/(2001)/may/05/fiction.reviews>.

Piatti, Barbara et al. (2008): "Mapping Literature: Towards a Geography of Fiction." Art & Cartography. University of Vienna, Austria. 31 Jan- 2 Feb. 2008. Paper.

——— and Lorenz Hurni (2009): "Mapping the Ontologically Unreal – Counterfactual Spaces in Literature and Cartography." *The Cartographic Journal. Art & Cartography Special Issue* 46.4: 333-42.

——— and Lorenz Hurni (2011): "Editorial. Cartographies of Fictional Worlds." *The Cartographic Journal. Cartographies of Fictional Worlds Special Issue* 48.4: 218–23.

Pile. Steve and Nigel Thrift (2000): *City A-Z: Urban Fragments.* London and New York: Routledge.

Pinder, David (2001): "Ghostly Footsteps: Voices, Memories and Walks in the City." *Ecumene, A Journal of Cultural geographies* 8.1: 1-19.

——— (2005a): *Visions of the City: Utopianism, Power and Politics in Twentieth Century.* Edinburgh: Edinburgh University Press.

——— (2005b): "Arts of Urban Exploration." *Cultural Geographies* 12.4: 383–411.

Pocock, David (1988): "Geography and Literature." *Progress in Human Geography*. 12.1: 87-102.

Porter, Eliot (1962): *In Wildness is the Preservation of the World.* San Fransisco: Sierra Club.

Preston, Chris (2003): *Grounding Knowledge: Environmental Philosophy, Epistemology, and Place.* Athens & London: The University of Georgia Press.

Price, Patricia (2013): "Place." *The Wiley-Blackwell Companion to Cultural Geography*. Eds. Nuala C. Johnson, Richard H. Schein, and Jamie Winders. Oxford: Wiley Blackwell. 118-29.

Pryce, Malcolm (2001): *Aberystwyth Mon Amour.* London: Bloomsbury.

——— (2003): *Last Tango in Aberystwyth.* London: Bloomsbury.

——— (2005): *The Unbearable Lightness of Being in Aberystwyth.* London: Bloomsbury.

——— (2007): *Don't Cry for Me Aberystwyth.* London: Bloomsbury.

——— (2009): *From Aberystwyth with Love*. London: Bloomsbury.

——— (2011): *The Day Aberystwyth Stood Still.* London: Bloomsbury.

——— (2013): "Mummy, It's Another of Those Horrid Emails from the King of Welsh Noir. *TheWordOnTheProm.com* 25 June 2013. <http://us7.campaign-archive1.com/?u=4f8b34b50b7f0f76a49dc8a21&id=e470e8d85d&e=5efe53c020>. See also <https://www.facebook.com/photo.php?fbid=49649790376326

3&set=a.496497887096598.1073741842.481805531899167&type=1&theater>.

——— n.d.: "I Have Not Now, Nor Ever Been a Librarian in Swansea." *MalcomPryce.com The contents of my bonce.* <http://malcolmpryce.com/intro.html>.

Pütz, Ole (2012): "From Non-Places to Non-Events: The Airport Security Checkpoint." *Journal of Contemporary Ethnography* 41.2: 154-88.

Ratzel, Friederich (1896): *Völkerkunde*. London: Macmillan.

Reed, Adam (2002): City of Details: Interpreting the Personality of London. *Journal of the Royal Anthropological Institute* 8.1: 127-41.

Reed-Danahay, Debodrah E. (1997): *Auto/Ethnography: Rewriting the Self and the Social*, Oxford: Berg.

Reijnders, Stijn (2010): Places of the Imagination: An Ethnography of the TV Detective Tour. *Cultural Geographies* 17.1: 37-52.

Relph, Edward (1976): *Place and Placelessness*. London: Pion.

Reuschel, Anne-Kathrin and Lorenz Hurni (2011): "Mapping Literature: Visualisation of Spatial Uncertainty in Fiction." *The Cartographic Journal Cartographies of Fictional Worlds - Special Issue* 48:4 293-308.

Rich, Adrienne (1986): *Notes Towards a Politics of Location*. New York: WW Norton.

Riley, Robert B. (1997): "The Visible, the Visual and the Vicarious: Questions About Vision, Landscape, and Experience." *Understanding Ordinary Landscapes.* Eds. Paul E. Groth and Todd W. Bessi, New Haven: Yale University Press. 99-110.

Ritzer, George, ed. (2002): *McDonaldisation: The Reader*. Thousand Oaks: Pine Forge Press.

Robson, Lloyd (1995): *Edge Territory*. Cardiff: Black Hat.

——— (1996): *Letter From Sissi*. Cardiff: Black Hat.

——— (2001): *Cardiff Cut*. Cardiff: Parthian Books.

——— (2003): "A Stain on the Map." *Cardiff Central. Ten Writers Return to the Welsh Capital*. Ed. Dannie Abse. Llandysul: Gomer. 95-104.

——— (2003): *Bbboing! And Associated Weirdness*. Swansea: Parthian.

——— (2005): "A Curious Pace." Finch & Davies 2005: 178-9.

Rose, Mitch (2002): "Landscape and Labyrinths." *Geoforum* 33: 455–67.

Rosenberg, Elissa (2012): "Walking in the City: Memory and Place." *The Journal of Architecture*. 17.1: 131-49.

Routledge, Paul (1996): "Third Space as Critical Engagement." *Antipode* 28.4: 399-419.

Sack, Robert (1997): *Homo Geographicus*. Baltimore: John Hopkins University.

——— (2004): "Place-Making and Time." *Reanimating places. Re-Materialising* (Cultural Geography Series) Ed. Tom Mels. Aldershot: Ashgate. 243-253.

Sadler, Simon (1998): *The Situationist City*. Cambridge: MIT Press.

Sauer, Carl (1925): *The Morphology of Landscape*. Berkeley: University of California press.

——— (1963): "Homestead and community on the middle border." *Land Use Policy in the United States*. Ed. Howard W. Ottoson. Nebraska: University of Nebraska Press.

Saunders, Angharad (2010): "Literary Geography: Reforging the Connections." *Progress in Human Geography* 34.4: 436-52.

Seamon, David and Robert Mugerauer, Eds. (1985): *Dwelling, Place and Environment: Toward a Phenomenology of Person and World.* Malabar: Kluwer.

Semple, Ellen (1911): *Influences of Geographic Environment*. New York: H. Holt & Co.

——— and Friedrich Ratzel (1941): *Influences of Geographic Environment*. Michigan: Constable & Company.

Sharma, Sarah (2009): "Baring Life and Lifestyle in the Non Place." *Cultural Studies* 23.1: 129-48.

Sharp, Joanne et al., Eds. (2000): *Entanglements of Power: Geographies of Domination/Resistance*. London and New York: Routledge.

Shaw, Ceri (2009): "An Interview with Niall Griffiths." *Welsh-American-Bookstore.com.* 31 Oct. 2013 <http://www.welsh-american-bookstore.com/index.php/Archive/interview-niall-griffiths-(2009).html>.

Sheers, Owen (2000): *The Blue Book*. Bridgend: Seren.

——— (2005): *Skirrid Hill*. Bridgend: Seren.

——— (2007): *Resistance. London*: Faber and Faber.

——— (2008): "Poetry and Place: Some Personal Reflections." *Geography* 93.3 Autumn: 172-5.

——— (2009): *A Poet's Guide to Britain*. London: Penguin.

——— (2011): "Hay Festival: When the Nazis Came to Hay." *TheTelegraph.co.uk.* 23 May 2011. <http://www.telegraph.co.uk/culture/hay-festival/8523902/Hay-Festival-When-the-Nazis-came-to-Hay.html>.

——— (2012): *The Gospel of Us*. Bridgend: Seren.

Sinclair, Iain (1971): *The Kodak Mantra Diaries*. London: Albion Village Press.

——— (1972): *Muscat's Würm*. London: Albion Village Press.

——— (1995): *Lud Heat and Suicide Bridge*. London: Vintage.

——— (1997): *Lights Out for the Territory*. London: Granta Books.

——— (2001): *Landor's Tower*. London: Granta.

——— (2003): *London Orbital*. London: Penguin.

——— (2005): "Crackerjack." Finch & Davies 2005: 189.

———, Ed. (2006): London: City of Disappearances. London: Penguin.

——— (2012a): "The Man in the Clearing. Iain Sinclair meets Gary Snyder." *London Review of Books*. 24 May 2012. 34.10: 35-38 <http://www.lrb.co.uk/v34/n10/iain-sinclair/the-man-in-the-clearing>.

——— (2012b): *Ghost Milk*. London: Penguin.

Smith, Ali (2004): "Life Beyond the M25." *TheGuardian.com.* 18 Dec. 2004 <http://www.theguardian.com/books/(2004)/dec/18/fiction.alismith>.

Smith, Phil (2011): "Mythogeography Works: Performing Multiplicity on Queen Street. Research in Drama Education." *The Journal of Applied Theatre and Performance*. 16.2: 265-78.

Snyder, Gary (1974): *Turtle Island*. New York: North Point Press.

——— (1983): *Axe Handles*. New York: New Directions.

——— (1990): *The Practice of the Wild*. New York: North Point Press.

——— (1995): *A Place in Space*. New York: North Point Press.

——— (2000): *The Gary Snyder Reader: Prose, Poetry and Translations, 1952-1998*. New York: Counterpoint Press.

Soja, Edward (1996): *Thirdspace: Journeys to Los Angeles and other Real-and-Imagined-Places*. Maldon: Blackwell.

——— (1999): "Thirdspace: Expanding the Scope of the Geographical Imagination." *Human Geography Today*. Eds. Doreen Massey, John Allen and Phil Sarre. Cambridge: Polity Press.

——— (2010): *Seeking Spatial Justice*. Minneapolis: University of Minnesota Press.

Solnit, Rebecca (2010): *Infinite City. A San Francisco Atlas*. Berkeley: University of California Press.

Spivak, Gayatri C. (1990): *The Post-Colonial Critic. Interviews, Strategies, Dialogues*. London and New York: Routledge.

Steinbeck, John (1945): *Cannery Row*. London and New York: Penguin.

Stern, Lawrence (2009): *Tristram Shandy*. London: Wordsworth Classics.

Tally, Robert (2011): "On Literary Cartography: Narrative as a Spatially Symbolic Act." *Nanocrit.com*. Jan. 2011. <http://www.nanocrit.com/issues/issue-1-navigation/literary-cartography-narrative-spatially-symbolic-act/>.

Thomas, Dylan (2003): [1954] *Under Milk Wood*. London: Phoenix.

Thoreau, Henry David (1862): "Walking." *Bartleby.com* n.d. <http://www.bartleby.com/28/15.html>.

——— (1894): *Walden*. New York: Riverside Press.

Thrift, Nigel (1996): *Spatial Formations*. London: Sage.

Tilley, Christopher (1994): *A Phenomenology of Landscape: Places, Paths and Monuments*. Oxford: Berg.

Torrance, Chris (1980): *The Cloud Book*. Neath: Cwm Nedd Press.

Tuan Yi-Fu (1974): *Topophilia: A Study of Environmental Perception, Attitudes, and Values*. Englewood: Prentice Hall.

——— (1977): *Space and Place: The Perspective of Experience*. Minneapolis: University of Minnesota Press.

——— (1996): "Space & Place: Humanistic Perspective." *Human Geography. An Essential Anthology*. Eds. John Agnew, David Livingstone and Alisdair Rogers. Oxford: Blackwell. 444-57.

——— (2004): "Sense of Place: Its Relationship to Self and Time." Mels 2004: 45-56.

Turner, Victor (1969): *The Ritual Process: Structure and Anti-Structure*. New York: Aldine.

——— (1974): *Dramas, Fields and Metaphors: Symbolic Action in Human Society*. New York: Cornell University Press.

——— (1982): "Liminal to Liminoid in Play, Flow, Ritual: An Essay in Comparative Symbology." *From Ritual to Theatre: The Human Seriousness of Play*. Ed. Victor Turner. New York: Performing Arts Journals Publications.

van Gennep, Arnold (1909) [1960]: *The Rites of Passage: A Classic Study of Cultural Celebration*. Chicago: The University of Chicago Press.

Vitterso, Joar, Mikal Vorkinn and Odd I. Vistad, O. (2001): "Congruence Between Recreational Mode and Actual Behaviour – A Prerequisite for Optimal Experiences?" Journal of Leisure Research 33.2: 137-59.

Wales Online (2011): "The Quality of Light is a Tale of Three Cities – Richard Collins." 23 Jul. 2011. <http://www.walesonline.co.uk/lifestyle/quality-light-tale-three-cities-1822793>.

Walford Davies, Damian (2012): Cartographies of Culture: New Geographies of Welsh Writing in English. Cardiff: University of Wales Press.

Warf, Barney and Santa Arias, Eds. (2008): *The Spatial Turn. Interdisciplinary Perspectives*. London: Routledge.

Warren, Kay B. (1997): "Narrating Cultural Resurgence." Reed-Danahay 1997: 22–46.

Weil, Simone (1955): *The Need for Roots. Paris: Gallimard.*

——— (2002): *Gravity and Grace*. London: Routledge.

Welsh, Irvine (1993): *Trainspotting*. London: Vintage.

Whatmore, Sarah (1999): "Culture-Nature." *Introducing Human Geographies*. Eds. Paul Cloke, Mike Crang and Mark Goodwin. London: Arnold. 4–11.

Whitman, Walt (2005): *Leaves of Grass*. New York: Signet.

Williams, Andrew Craig (2001): "The Days of Our Lives." *New Welsh Review* 54: 79-80.

——— (2005): "Rosie's Last Week." Finch & Davies 2005: 202-10.

——— (2011): "Cardiff and Its People Have Shaped Who I Am Today." *WeAreCardiff.co.uk*. 19 Oct. 2011. <http://weare cardiff.wordpress.com/(2011)/10/19/cardiff-and-its-people-have-shaped-who-i-am-today-andrew/>.

Williams, Carol (2002): *Sugar & Slate*. Aberystwyth: Planet.

Williams, Gwyn (1977): *The Land Remembers*. London: Faber & Faber.

Williams, John (2004): *Michael X*. London: Century.

——— (2005): "The Legend of Tiger Bay." Finch & Davies 2005: 217-234.

——— (2006): *The Cardiff Trilogy*. London: Bloomsbury.

Young, Iris Marion (2000): *Inclusion and democracy*. Oxford: Oxford University Press.

Appeared earlier in the SPATIAL PRACTICES:
AN INTERDISCIPLINARY SERIES IN CULTURAL HISTORY, GEOGRAPHY AND LITERATURE

Navigating Cultural Spaces: Maritime Places
Volume 18 – *Edited by Anna-Margaretha Horatschek, Yvonne Rosenberg and Daniel Schäbler*

Perspectives on Mobility
Volume 17 – *Edited by Ingo Berensmeyer and Christoph Ehland*

Cormac McCarthy and the Writing of American Spaces
Volume 16 – *By Andrew Keller Estes*

Placing Poetry
Volume 15 – *Edited by Ian Davidson and Zoë Skoulding*

Topodynamics of Arrival: Essays on Self and Pilgrimage
Volume 14 – *Edited by Gert Hofmann and Snježana Zoric*

Land & Identity: Theory, Memory, and Practice
Volume 13 – *Edited by Christine Berberich, Neil Campbell and Robert Hudson*

No Maps for These Territories
Cities, Spaces, and Archaeologies of the Future in William Gibson
Volume 12 – *By Karin Hoepker*

Reflective Landscapes of the Anglophone Countries
Volume 11 – *Edited by Pascale Guibert*

Process: Landscape and Text
Volume 10 – *Edited by Catherine Brace and Adeline Johns-Putra*

Futurescapes: Space in Utopian and Science Fiction Discourses
Volume 9 – *Edited by Ralph Pordzik*

More titles: www.rodopi.nl